Dogecoin Tricks
101 Cool Things to do with the Original Memecoin

chromatic

Dogecoin Tricks
101 Cool Things to do with the Original Memecoin

Editor: chromatic
Logo design: Devin Muldoon
Cover design: Natalie Wood, https://www.instagram.com/neverwood.art/

ISBN-13: 979-8-9898637-0-9

Published by Onyx Neon Press, http://www.onyxneon.com/. The Onyx Neon logo is a trademark of Onyx Neon, Inc.

Onyx Neon typesets books with free software, especially Linux, Perl, PseudoPod, and LaTeX. Many thanks to the contributors who make these and other projects possible.

First edition December 2024, second printing June 2025.

For more details, see https://ifdogethenwow.com/books/dogecoin-tricks/. Please *do* share this link with your friends and colleagues.

Thanks for reading!

For Rosie
April 26, 2024
8:52 pm

Contents

Foreword

As I write this foreword to chromatic's beautiful book, in the corner of an empty well-lighted bar in the middle of the night, cannoli waiting for me in a tray next to my computer, my mind wanders. It goes back to those early days of December 2013, when I was still living in a small apartment overlooking the river Tejo in Lisbon. With my partner, I had finally decided to understand more about Bitcoin and cryptocurrencies. The idea of trying to use my GPU to mine a cryptocurrency seemed a good point of entry.

Back then, it was already impossible to mine Bitcoin with a GPU. Litecoin was a possibility. But then we came across the first posts about Dogecoin on Twitter and Reddit. The "fundamentals" that attracted us to Dogecoin were the yellow dog in the logo and the Comic Sans font. Yes, it was clear this was a fork of coins derived from Bitcoin, but the joke of it all was so funny and endearing that all the rest did not matter. We looked at our dog K., back then still young and sleeping at our feet, and the decision was taken. I can still hear the whirring sound of the GPU in that small one-room apartment. It would serve as a background to my work, sleep, and indoor soccer games with K. for a few months.

Since 2013, reports of Dogecoin's death have been greatly exaggerated. The coin and the community around it that was supposed to die off and disappear proved to be incredibly resilient. A testament to its decentralized and permissionless nature, to its community of wacky heroes and occasional villains, but also to the power of memes and jokes, the power of dogs and the Comic Sans font. "A laugh that will bury them all," is a phrase that keeps ringing in my head whenever I see Dogecoin in the wild.

More than 10 years have passed, and while cryptocurrencies have evolved, education around them is still lacking and is often focusing on the speculative side of it all. Everything is about the market, instead than on how the tech works, why it works, and why permissionless decentralized money can matter.

Because Dogecoin is, for the most part, a fork of Bitcoin with the good heart of a

i

dog. The technical foundations are solid–insofar as Dogecoin is well maintained and its network is healthy–but the face shown to the public, the ubiquity of the meme, its irony and friendliness, and the often welcoming community make it a favorite point of entry for those new to crypto. Dogecoin holds this special position of being the gateway to crypto for so many, a special vantage point that should come with a certain sense of responsibility.

This is why a book like this one is so important. chromatic, who also contributes as a developer and maintainer for Dogecoin Core (Dogecoin's reference implementation), guides you through the concepts behind cryptocurrencies with poise and panache. The brilliant chapters opening the book introduce important yet potentially unintuitive concepts in a very approachable and rewarding way. Through the rest of the book, chromatic then uses dog money as a playground to teach concepts that transcend it.

A writer and free software programmer with a profound love for humanities in his heart, he might demystify "mnemonic passphrases" by using the words as solutions to riddles in a treasure hunt, in a chapter filled with Easter eggs and references to the best adventure game ever. He spends time to teach you how to run a Dogecoin Node and to monitor and interact with it, but he also needs to show you how to turn your transaction IDs into music, how to create a Dogecoin-powered jukebox, how to hide a wallet address in the photo of a lovely pug, how to tip a Geocache with Dogecoin. Each one of these tricks introduces or reinforces important concepts by asking you to play with Dogecoin and to build small things. "Learning by playing" is a way of teaching that seems to capture Dogecoin's essence just perfectly.

chromatic writes with lightness and an ebullient sense of fun, but without eschewing responsibility. He constantly reminds the reader of risks and possible pitfalls behind many of these tricks. The "Understand the Risks" section at the end of his tricks are invaluable.

A chapter on how to "Manage a Dogecoin Arcade" completes this book, tying together everything chromatic has taught you. The chapter introduces design ideas that might very well power other ventures–such as a cannoli smuggling device–and like many other chapters of the book tries to bring Dogecoin into the real world.

Because in the end, I believe that a cryptocurrency has real value only when people use and transact with it. A cryptocurrency lives when it serves the purpose of sending money from point A to point B without a centralized authority behind it, and when it does so safely and reliably. Dogecoin can do this with the lightness of a joke, the irony of the dog reminding you that money has worth because people have agreed to transact with it, because they have decided it stands for something valuable. Dogecoin reminds you that people should have power over money and not the other way around.

And with his wit, warmth, and intelligence, chromatic manages to introduce concepts and ideas that will serve his readers well, beyond Dogecoin.

I have indirectly known chromatic for a long time through the legendary Perl books he wrote, and which I pillaged in my career as a web developer for desperados and cannoli smugglers running at the edges of this troubled world.

This book has already proven to be just as useful as his earlier books. Have fun with it.

Mishaboar, April 2024

Preface

Dogecoin is the Internet's original meme-based cryptocurrency.

That sentence may read like a bunch of nerdy nonsense. Don't worry; it is! Fortunately, this book will make it clearer. While it will remain nerdy nonsense, you'll understand it and be able to have fun, do cool things, and show off your new knowledge to friends at parties[1].

Let's explain that first sentence.

Dogecoin is the subject of this book.

The Internet is a globally-distributed network of computers where people and machines communicate in words, pictures, video, audio, and other data.

A *meme* is an idea, often clever or ironic or sarcastic or joking, which spreads easily between people. It often takes the form of an image or video.

Cryptocurrency is a mechanism to record digital transactions over the Internet between people who don't necessarily trust each other as individuals but who agree, as a group, to communicate in rigorous ways to make the entire network of people trustworthy enough to serve as a mechanism of financial exchange.

Whew.

The crypto stuff can get a lot deeper, but most of it summarizes to this: you don't have to take someone's word that they're trustworthy. You should be able to verify it.

If that sounds like a lot of work with a bunch of serious nerds doing serious nerd stuff, remember that Dogecoin has a friendly dog mascot with a bunch of silly memes. The point is to have fun.

Fun is what this book is about.

---

[1]Talk about other things too; we want you invited *back*.

What Do You Need?

This book assumes you have access to a computer somewhere (though some of the tricks will work on mobile devices such as phones and tablets), have access to the Internet, and can download and install and run software.

As of the time of writing, you can visit dogecoin.com for links and information about how to download and run the Dogecoin Core. Don't take our word for it though; check online if that's still the home of the Core. Test that against other sites such as https://github.com/dogecoin/dogecoin, the Dogecoin Reddit at https://reddit.com/r/dogecoin/ and other locations. If something seems off to you, then ask questions. Honest, helpful people will give you good answers and help you verify them for yourself.

How to Read This Book

Fun is the serious business of Dogecoin, so read this book in any way that *you* enjoy. Maybe that's front to back or back to front or skipping around. Maybe you read a couple of pages before going to sleep to relax you or teach your brain a lesson. The only *wrong* way to read this book is a way that you don't enjoy.

The book starts by assuming you're new to cryptography, cryptocurrency, and Dogecoin. Earlier pages introduce concepts that later pages expand. Skim the table of contents. Look at the names of every chapter and every tip. If something grabs your attention, read that first. Repeat.

This book also makes explicit links between ideas. Concepts and themes repeat throughout the book–on purpose. It's just as valid to breeze through this book on a lazy summer afternoon as it is to skim through it over the period of weeks, months, or years. Feel free to set it down and come back to it later; all of the information will be waiting for you. Ideally each tip has enough breadcrumbs to point to other places that you can remind yourself where you were in your journey at the time.

Put this book on your bedside table. Face it out from your bookshelf during video calls. Carry it while walking, riding the train, or enjoying a cannolo on a park bench downtown. Lend a copy to your friends and donate a copy to your local library. Above all, embrace the fun. Being in control of your own coins is liberating and powerful. Learning can be fun and playful.

With that said, take note of a couple of things.

Cryptocurrency Isn't Only Dogecoin

Although this book *focuses* on Dogecoin, many concepts and ideas apply to multiple cryptocurrencies, especially coins with a similar lineage. Many examples will work with small or minor modifications. If you prefer another coin over Dogecoin, that's totally okay! You can still learn a lot and have the appropriate amount of fun. We all share some common ground, including code, so we can all learn from each other.

What will you need to change? The names of executable files, for example, are different between coins. You launch the Dogecoin Core wallet GUI via a program named dogecoin-qt and the Pepecoin equivalent via pepecoin-qt. Similarly, wallet addresses are all different, as are the ports used to connect to each coin's network and RPC mechanisms. It's okay if this detail seems overwhelming now; it will make more sense when you really need to know it.

Consult your coin's documentation for details on what needs to change. This includes Bitcoin[2], Litecoin[3], and related forks and spinoffs, including Pepecoin[4] and Dingocoin[5]–even the immediate predecessor of Dogecoin, Luckycoin[6].

Programmable Money Brings Risks

Even if the only thing we used cryptocurrency for were to track scores in a massive, multiplayer online game, we'd still face risk. Money or any value makes those risks more serious. Scammers, thieves, grifters, and other bad actors are, unfortunately, a part of the landscape. Where possible, this book attempts to identify risks to you, the people you care about, and the other people in the community. It's up to you to keep yourself and others safe, but it's only possible to ameliorate risks you know about. Read the risk sections carefully and think about other things that could go wrong.

If you're not sure, ask. After you ask, verify. This is also a book about ways to prove things, and knowledge and proof are power.

---

[2] See https://bitcoin.org/.

[3] See https://litecoin.org/.

[4] See https://pepecoin.org/.

[5] See https://dingocoin.com/.

[6] See https://luckycoinfoundation.org/.

You Can Grow with This Book

Perhaps you picked up this book because the cover has cute dogs. Though you're not yet ready to build a pinball museum where people pay with cryptocurrency, that's okay! If you read one tip every week, you'll get through this book in two years of lazy Sunday afternoons. At that point, you may be full of other ideas no one has thought of before as well as new skills to bring them to life.

This book contains code in languages including Python, Ruby, JavaScript, Bash, and Perl. If you're not a programmer, that's okay! The code will explain and illustrate concepts, but you don't have to program to *use* programmable money.

With that said, you *will* get the most out of this book if you're comfortable installing and configuring software, especially working with configuration files and the command line. With help from a trusted friend, you can learn a lot and take control of your system in ways you previously didn't understand.

In short, if you're comfortable working with a spreadsheet, you know enough to get technical with Dogecoin!

Credits

This work, as expected, has benefited from much wonderful assistance.

Thank you to RNB[7] and Rachel[8] for feedback which led to several tips!

Thank you to Mishaboar[9] for advice, perspective, and helping promote a deep empathy for other people–as well as the lovely foreword!

Thank you to Patrick Lodder[10] for technical advice and trenchant corrections.

Thank you to Brett Warden[11] for electronics and wiring advice, and the Dogecoin Pinball switch schematic.

Thank you to @Pepecoininfo[12] for reporting several typos.

Any remaining errors are the fault of the sometimes-too-clever author[13].

---

[7] https://x.com/RNB333

[8] https://x.com/meta_rach

[9] https://x.com/Mishaboar

[10] https://github.com/patricklodder/

[11] https://x.com/bwarden

[12] https://x.com/pepecoininfo

[13] https://x.com/chromatic_x

## CHAPTER 1

Understanding Cryptography

Wow, that sounds yawn-inducing!

Cryptography is fundamental to cryptocurrency. That's what "crypto" in the name means[1]. While you don't have to understand all of the deeply mathematical details of elliptic curves, public/private key ciphers, and Merkle trees, knowing the basics of *how* everything works and fits together will help you be safer and more secure as well as have more fun learning about Dogecoin and cryptocurrency.

You'll need a couple of sharp pencils, some paper, and a hand calculator–or a good spreadsheet, if you're comfortable with that. Set aside some space to work and keep an open mind.

While this chapter intends to explain the fundamentals of cryptography, it's not going to make a rigorous, mathematically-sound attempt. It's going to use analogy and example so you'll get a feel for how things fit together. This won't provide the same security as things cryptographers do all day every day; it's intended to explain and entertain, not secure.

Tip #1   Make a Hash of Fingerprints

Every concept in the cryptographic world exists for at least one reason, and most of those reasons are, at their core, "don't trust anyone" and "verify everything".

For example, suppose you want to download this book from `ifdogethenwow.com`. How do you know the PDF you're downloading is the real PDF and not a counterfeit someone maliciously pointed you at instead?

What if there were a quick and easy way you could have your computer scan the *contents* of the PDF and give you a code you could compare with the code your

---

[1]Crypto as in "hidden", not "graveyards", but if you want to be spooky and do Dogecoin, that's cool too!

author provided? If your author published that code (and the way to calculate it) on GitHub, Twitter, Reddit, and any other place and all of those codes matched *and* you calculated that code yourself, how much more confident would you feel that the PDF you downloaded matched exactly the one your author produced?

Cryptographic Hashes

That's the idea behind cryptographic hashes. Given a well-understood way to look at every piece of data in a file (whether it's a text file, a song, a program, an image, whatever), can you produce an unpredictable number that would be difficult to forge?

Warning: The explanation you're about to read omits a lot of important details in favor of simplicity. The goal is to allow you to calculate things by hand so you understand *how* the approach works, not to teach you all of the details that make it secure. Do not use this for any purpose where actual security matters.

Simple Hash Math

Let's try this by example. Grab a piece of paper and write out the alphabet, A to Z. Beside each letter, write a number. A is 1, B is 2, all the way to Z is 26. This gives you a way to turn letters to numbers.

Pretend your name is Devlin Telshire. Now write your first name. Space it out. Beneath each letter, write the number corresponding to each letter. D is 4, E is 5, all the way to E is 5 again.

Now add up all the numbers. You should get 66. That's not bad. That's a fingerprint of "Devlin", but it's not a good fingerprint. If you pick a different name, how easy is is to get the same value? The day of the first writing of this tip happened to be the Ides of March, so it's a nice coincidence that the name Brute also adds up to 66. That collision isn't good.

Reducing Collisions

If this book PDF only contained the word "Devlin", someone could substitute the word "Brute" and the resulting check would be the same. To avoid forgeries, we want the resulting code to be much much more difficult to guess or perturb. That means we need bigger numbers, and we want them to vary much more when things change.

If you look closely, you'll notice that Devlin will produce 66 and Brute will produce 66 but so will Evlind and Vlinde and Ruteb and Uterb. Even though these aren't common names, they're produce the same results. How about instead of adding numbers together, we multiply the number by its position in the text?

D has a value of 4 and is the first letter, so multiply 4 by 1 to get 4. E has a value of 5 and is the second letter, so multiply 5 by 2 to get 10, and so on. You should get a value of 257 for Devlin. You should get 206 for Brute. Even better, you'll get 215 for Evlind and 150 for Ruteb. That's more like it.

Handling Larger Data

What happens when you want to get a fingerprint for an entire sentence? A paragraph? A book? These numbers will get prohibitively large quickly. For example, the value for "In the beginning the Universe was created. This has made a lot of people very angry and been widely regarded as a bad move." is over 49,000. That gives an attacker a lot of opportunities to find a lot of text that could add up to that number (find the payload you want, then keep padding it with "A" until you hit your number).

How do you secure the hash against that?

Think for a moment about what makes a random number truly random. It's unpredictable. You can't tell how it was calculated. If there's an *algorithm*[2] or a formula you could use to generate it, it's not obvious, no matter how hard you work, how to force an output you want. The field of information theory uses a term called *entropy* to refer to the apparent available randomness of a system (see Embrace Entropy, pp. 21 for more details).

Small numbers like 49,000 don't have much entropy available[3]. What if the number were 10 times larger? 100? 1000?

One way to make adding the letter "A" less useful to forgers is to start with a much larger number. What happens if, rather than starting from zero, every time you calculate this hash you start with the number 12,345,678? In that case, Devlin hashes to 12,345,935 and Brute to 12,345,884. For the Douglas Adams quote, the value is 12,394,893. This works better for larger data than smaller data.

Is there a way to ensure we always have a fixed amount of entropy, regardless of the size of input data?

Clocking Out

These numbers in the range of 12 million have about 24 bits in them. That's not a lot in the grand scheme of things, but it gives us an opportunity. We know that

---

[2]This is a fancy computer science word meaning "Repeatable mathematical process". Think of it like a recipe for doing math, but unlike a cookie recipe you shouldn't skip steps, for example to leave out the raisins.

[3]Convert 49,000 to binary and count the digits. In this case, you need 16 bits to represent the number.

every number is going to be larger than 12,345,678. We know that a small data set will be close to that number, while a large data set will hash to a value much larger than that number.

Now image that, at every calculation, the result ought to be larger than 10,000,000 but less than 49,876,543 (see Roll Over Your Odometer, pp. 5)? In other words, while adding the value of a letter's position times its alphabet value, you must always add 10,000,000 and always subtract 49,876,543 repeatedly until you get a value within the expected range?

Brute produces 12,469,341. Devlin produces 22,469,392. The quote produces 44,740,576. That's better; the length doesn't make numbers appreciably smaller or larger.

Also note what happens if you append A to each of these. Brutea makes 22,469,347. Devlina makes 32,469,399. The quote makes 44,740,576. That's a lot less predictable than before. Now an attacker who wants you to download a malicious file that seems like it's exactly the same file your author provided has to do more work to fool you.

What's Really Going On

This explanation omits a lot of important details. Do not use this approach for anything where security matters. You can use this for fun to understand some of the ideas yourself or to share with others the basic concepts, but there are trivial attacks on this approach that any knowledgeable and determined attacker could exploit with very little work.

There's a reason cryptography is difficult. Leave devising these algorithms to professionals who practice rigorous, public peer-review.

Perhaps the best and most thorough explanation is Bruce Schneier's book *Applied*

Cryptography[4]. While this chapter attempts to explain things by analogy and example, Schneier explains how things *actually* work and doesn't shy away from important details that make systems *actually* secure.

What would make this algorithm better? More bits of entropy; 4 - 24 bits as shown here are easier for you to calculate, but easier for you to calculate by hand is much, much easier for a computer to calculate. In general, the larger the numbers, the more work a computer has to do to try all of the possibilities that could make a collision. This is an important fact to remember throughout the book.

While this hashing algorithm might work okay for a very simple data structure used in a programming language, a really good cryptographically secure hashing algorithm will do a lot of work to make sure that every change to the document–every character or byte–can affect as many of the bits in the output as possible, in as unpredictable a way as possible. This algorithm doesn't do that, but hopefully it's shown you enough of the theory behind it that you can understand why that's important and how a good algorithm might work.

  # Tip #2   Roll Over Your Odometer

Imagine it's 1979 and you're really, really good at the arcade game Asteroids. Every 10,000 points you gain a new ship. What's the high score?

99,990.

At that point, the score resets to zero, although apparently you can buy a special add-on which increases this limit to 9,999,990[5]. This mod still has an upper limit, at which point the score resets to zero and starts over. The previous sentence has a pun, and you're about to understand why.

This feature turns out to be really important in cryptography.

Romulus, Remus, Modulus, and Remainder

Think of the high score counter as an odometer in a vehicle or the hands on a clock. There's an upper limit (12 or 24 hours on a clock, hundreds of thousands of miles or kilometers in an automobile, a hundred thousand or a hundred million points in Asteroids). Once you exceed the limit, you start over.

---

[4]See https://amzn.to/42pyFeS.

[5]The things you learn researching cryptography for a book about funny dog money! Search for "Asteroids High Score Kit".

This is an expression of a mathematical concept called a *modulus*. You may have heard of it as a *remainder*. You can even demonstrate it by counting on your fingers, from one to ten.

Suppose you want to add the two prime numbers five and seven. Count on your fingers. Raise seven fingers, then raise five more. When all of your fingers are up[6], lower them all again but keep counting. The number of fingers you have raised at the end is 7 + 5 mod the number of fingers. If you have 10 fingers, 7 + 5 mod 10 is 2. If you have 9 fingers, 7 + 5 mod 9 is 3.

What's Really Going On

Why does this matter? If you're doing something like calculating a hash of data (Make a Hash of Fingerprints, pp. 1), you want the output to have two properties. First, you want the output to have the same number of bits regardless of the size of the input. Short inputs and large inputs should produce output that's the same size. Second, you want attackers to have a high difficulty predicting what kind of changes to inputs produce what kind of outputs.

In both cases, dealing with really large numbers but constraining the results to a narrow range helps.

That's what modulus does. You'll see it in all sorts of cryptographic applications. It happens in non-cryptographic applications too, including days of the week, minutes in the hour, degrees of a circle, months of the year, names of notes in the 12-tone Western scale, and the Roman nundinae, or 8-day market/work week.

For our purposes, it has one other essential property. Given arbitrary inputs and a modulus, you actually *lose* information. Multiply 2 by 7 mod 10 and you get 4. Also multiply 6 by 4 mod 10 and get 4. Multiply 12 by 2 mod 10 and get 4. Given the answer of "4" and a modulus of "10", you can only guess at many, many integers you could multiply together to get that answer.

In other words, you're going to have a lot more difficulty *reversing* the calculation to figure out its initial inputs, even if you have part of the equation. That irreversibility turns out to be an essential property of cryptography.

---

[6]If you have more or fewer than ten fingers; that's okay. You'll get a different answer, but the same concept applies!

Tip #3   Forge a Chain

Isn't it interesting how the word "forge" in English can mean either "create something successful and strong" or "produce something for purposes of a deception"?

Imagine you're trying to keep a group of rambunctious young goats closed up in their pen to enjoy the sunshine (and out of the garden, because they will eat your baby tomato plants in the blink of an eye). You want a strong fence with a gate, and you need a strong chain to keep that gate closed. In a chain, every link connects to another link. The strength of every part of the chain depends on the strength of every link in the chain.

That metaphor applies to your bank account too. If you're not sure about the transactions from a month ago adding up correctly, how can you trust that the transactions of yesterday and the balance of today make any sense?

Given that this is a chapter of a book about cryptocurrency, you've probably already guessed that cryptography has a few ideas on how to answer that question.

Validate Data

Assume you have a niece or nephew who wants to babysit your goats to earn some spending cash. We'll call this kid Kid A, because that's a nice gender-neutral name and baby goats are also called kids. You don't have a lot of time to keep records[7], and Kid A is really bad about writing down their time, so you want to avoid arguing over when you paid and how much. You need to track this in a way that doesn't mean you have to go back into your phone's text messages every couple of weeks to reconcile things.

You can fix this! First, agree on the format of an invoice. Kid A will report their hours in the form:

```
Goat Sitting
March 5 - 18 2023
12 hours
```

You will report in the form:

```
Goat Sitting Payment
March 5 - 18 2023
12 hours
1200 Dogecoin
```

---

[7]Or, in the case of Kid A, you prefer streaming individual songs.

That gives you the base data format you can agree on. If you use a cryptographic hash (Make a Hash of Fingerprints, pp. 1) on each set of data, you can get a unique fingerprint to prove no one tampered with the payment request/report forms. Use SHA-256 hashing, for example, and you'll get hashes like `24a0ae...` and `84ab55...`. If Kid A provides the hash of their invoice and you provide the hash of your payment, you have a record you can keep in multiple places.

That can help reduce the possibility of accidental or deliberate tampering, but validation of invoicing and payment is a different story.

Even better, let Kid A keep their own hash of their invoice and you keep your own hash of your payment and use them to validate each other. When you receive an invoice, calculate your own hash. When Kid A receives a payment, let them calculate their own hash. This can work together really well.

Validate and Chain Data

What if the validation for one invoice depended on the previous invoice? For example, if March 5 - 18 is the first invoice, you know the hash of that. What if you add the hash of the previous form as input in the next form? For example, the payment for the invoice could change to be:

```
Goat Sitting Payment
March 5 - 18 2023
12 hours
1200 Dogecoin
Follows: 24a0ae...
```

Now the hash of the payment becomes `eadfd8...`. Add that to the next invoice:

```
Goat Sitting
March 19 - April 2 2023
20 hours
Follows: eadfd8...
```

...and the new hash is `fd13b3...`. If you keep this going, you can validate two things. First, that no one has tampered with any individual invoice or payment. Second, that no one has tampered with the history and lineage of the invoices or payments.

In this mechanism, validation is easy. If you've both kept your own hashes when you've submitted your forms to each other, you can look at the response message and check against your own hash. If anything's different, raise a red flag.

What's Really Going On

As it turns out, that latter property is essential to all sorts of applications, including cryptocurrency and managing the source code for computer programs. By publishing the history of changes, or even only the hashes for each piece of data, anyone with access to the data can verify that every piece of data is in the correct place and everything before and after it belongs where it is.

If you dig into computer data structures or cryptography, you may hear this referred to as a *Merkle Tree* or *Merkle Chain*.

In this payment example, you don't even have to validate the entire chain of invoices and payments to figure out if something went wrong; if you've been validating as you go, all you have to do is validate the most recent message and you can be confident everything else before now is still untampered and accurate and pristine. You can–and sometimes *should*–still go back throughout history and prove the entire chain is valid, but you don't have to assume it is and you don't have to verify everything every time you make a change.

  ## Tip #4   Shorten Large Numbers

So far, this chapter has discussed some complicated concepts without getting to deep into the details of cryptography. It's time to go deeper, and in cryptographic terms that usually means handling large numbers.

How large are large numbers? When making cryptographic hashes of data (Make a Hash of Fingerprints, pp. 1), even the simple examples has to use large numbers to get any sort of security at all–and the security of the example is still very low. One cryptographically secure hashing algorithm, SHA-256, has 256 bits of security–256 individual units that can be either 0 or 1. If you were to write out a 256-bit number in base 10, you'd have to write 78 digits. That's big.

If you had to write out these numbers by hand on a regular basis, you'd get tired. Worse, the larger the number, the more chances you might transpose two digits or get something wrong or leave something out. To make cryptography useful, we have to make it easier to work with large numbers.

Tens without Ten

One solution hides behind the idea of modulus math (Roll Over Your Odometer, pp. 5). Think what would happen if everyone had nine fingers. You can count all the way up to nine, but to represent the idea of ten you'd have a representation something like "one full set of numbers and one more", or 11 base 9^8. If you have ten fingers, tape your pinky and ring finger together on your left hand and try counting to twenty. You'll have to represent this as 22 base 9.

If this still seems weird, think of it this way. Take the tens position (the first 2 in 22) and multiply it by nine. That gives eighteen. Now add the ones position (the second 2 in 22), and you get twenty. 20 base 9 plus 2 base 9 is 2 * 9 with 2 as the remainder.

If you counted on your fingers and toes (and have ten of each), you could count to twenty and represent that as 10 base 20. If that makes sense, you can probably see the trick already.

Counting to Ten with the Alphabet

Counting to twenty in base 20 is weird though. What do you use to represent eleven? 11 in base 20 is the value twenty-one, so we need another digit or character or perhaps even a letter.

Hang around techies for long enough and you'll hear some nerdy talk about something called *hexadecimal* notation. That means 16. Hexadecimal notation is a way of representing numbers in base 16, so the value of 10 base 16 is sixteen and the value of 20 base 16 is thirty-two and the value of 64 base 16 is one hundred.

Already at that point—one hundred—the hexadecimal notation is one digit shorter than standard base 10 notation. A SHA-256 value which would normally take 78 digits to represent in base 10 needs only 64 digits in hex.

What does ten look like? In base 16, that's A. Eleven is B. Twelve, through and fifteen are C, D, E, and F. You already know what sixteen is. Seventeen is 11. Twenty-six is 1A (read it as one times sixteen plus ten). Beyond that, use a hexadecimal calculator; don't worry about doing this in your head.

With 26 letters in the boring, non-accented Roman alphabet, you can represent values up to base 36 with the digits 0 through 9 and the letters A through Z. Base 36 would make the large cryptographic hashes we're dealing with much, much shorter—but we can do better.

Fifty-Eight is Larger than Thirty-Six

What if we added lower-case letters? We could represent thirty-seven with a, thirty-eight with b, and so on.

---

[8]Base *n* means you're using *n* fingers. It's 9 in this example, but it could be 10 or 16 or 58.

The only problem with this approach is that an upper-case letter O[9] can look a lot like the digit zero O, and a lower-case letter l[10] can look a lot like the digit one 1 or the upper-case letter I[11]. This can be a problem for really large numbers, especially if people are going to write them down *or* if there's a chance that malicious actors will try to fool other people to get control of their data or scam away their hard-earned Dogecoin.

Litecoin, Bitcoin, Dogecoin, and other cryptocurrencies use a format called Base-58. The allowed characters are the digits 0 through 9, the upper-case letters A through Z, and the lower-case letters a through z *except* for 0, O, I, and l. A Dogecoin address or key is pretty long even in this format, but it's a *lot* shorter than if you had to write it out in base 10.

What's Really Going On

Base-58 achieves two goals. First, it shortens large numbers by using a larger set of digits in the representation. Second, it avoids characters that could be ambiguous. Furthermore, the idea of attacks (or typos or transcription errors) against addresses or keys represented in Base-58 has led to the development of something called Base58Check[12], which is Base-58 with a couple of extra digits that provide a checksum of the data. This helps identify any errors and can, in some cases, help recover from errors.

While it's may be useful to remember specific values in base 16 if you're a programmer, you usually don't have to worry about converting between bases. Use a calculator if you need to (or rely on trusted software to do it for you). You don't have to be able to perform the conversion in your head as long as you understand why this exists and, at a basic level, how the math works.

  Tip #5   Host a Puzzle Race

Imagine it's January 2003. You're trying to finish graduate school. To relax on the weekends, you and your friends get together to make and eat soup and play some kind of game. Mario Kart Double Dash won't release for another 10 months, and

---

[9] O for Oscar.

[10] L for Lima.

[11] I for Indigo.

[12] See https://en.bitcoin.it/wiki/Base58Check_encoding.

none of you speak enough German to play the *good* European board games[13].
What can you do to pass the time?

How about a math game?

What Makes a Game Fun?

A math game? Why would anyone *play* this[14]? Why does anyone play any game?

- It offers a challenge to overcome. The end isn't preordained. You could win or lose.

- It has fair rules. Everyone understands the rules and everyone agrees on them, so you're all playing the same game.

- Everyone has a chance to win. (Okay, if you're playing chess or tennis against a world champion, maybe you're doing it for the story you can eventually tell, but playing a game you know you'll always lose quickly stops being fun.)

- It's rewarding. Maybe you get to choose one soup to keep next week and one soup to skip. Yay for beer cheese, boo for clam bullion. There's always some kind of reward, even if it's only bragging rights for the week.

These aren't the only ingredients for fun, but if none of these ingredients are present, you might as well get back to writing your thesis. How does this make a puzzle then?

Race to (Some) Target

A cryptographic hash (Make a Hash of Fingerprints, pp. 1) is, in theory, so difficult to reverse-engineer that you'll spend uncountable amounts of time trying to find two inputs that produce the same output. That principle holds true, but it holds more true the longer the output is and less true the shorter the output is.

If the output hash had only one bit of data–if the outputs were only 0 or 1–then it would be trivial to guess at an input that would produce the same output as another input. With a good hash algorithm, you'd have a 50% chance of being right with any one guess.

---

[13] Sure, there's Monopoly (Chisinau edition), Rummicluj, and Brussels Candyland, but....

[14] At least when they could roll dice and pretend to buy a giant concrete building in Moldova instead.

12

If the output hash had *two* bits of data–if the outputs were 0, 1, 2, and 3–you'd have a 25% chance of being right with any one guess. With three bits of data, you have a 12.5% chance of being right with any one guess, and so on.

That's true of the length of the output hash. It's true of other properties as well. Hold that thought for a moment.

What are the rules of this game? Suppose they are:

- Agree on a hashing algorithm

- Agree on a starting input

- Everyone pick a random number to combine with the input

- Everyone produce their new hash

- See if anyone has won

- Repeat from picking random numbers

Everyone starts with the same partial input, combines that with their own unique piece of input, and checks the output. If anyone has hit the target, they win! If not, pick a new unique input and check again!

Suppose you and your friends all agree on a hashing algorithm that produces a six-digit hexadecimal number (Shorten Large Numbers, pp. 9) between 0x100000 and 0xffffff. This range gives you 15.7 million possible numbers. Your soup will get cold before anyone guesses the right number, on average.

Race to a Pattern

What if you weren't looking for a specific number, but instead a number that matched a pattern?

Suppose the first person to find a hash starting with 1 could win. There are sixteen digits in hexadecimal (and, because the range of outputs must be greater than 1000000, only the digits 1 through 9 and A through F can be in the first position), so if the hash distributes its outputs evenly, everyone has a one in fifteen or 6.67% chance of finding a matching output with any single guess. With five friends, you might have a winner in three rounds.

What if that's not difficult enough, because one night everyone brings their laptop to soup night and calculates and checks their inputs a billion times faster? What if you increase the prefix to 12? There are fifteen possibilities for the first digit and sixteen (0 is allowed) for the second, so there's a one in fifteen times sixteen (one in 240) chance to get a matching output.

The more specific you make the prefix, the lower the chance of any one single guess matching the pattern, *assuming the hash produces outputs distributed across the range of allowed values evenly*. If there's a flaw in the hash or the algorithm, someone can exploit it. For something like soup night, that might not matter if you invite only people you trust. For a global cryptocurrency with billions of coins at stake, it's important to use an algorithm that's been reviewed by many, many serious and careful researchers.

What's Really Going On

Dogecoin miners don't get together to eat soup, at least not as a matter of network protocol. They *do* race to solve a puzzle. Everything in this game about picking a random number to hash together with a specific value to get an output matching a pattern is true, or at least true-ish.

In Dogecoin, Bitcoin, Litecoin, and other proof-of-work cryptocurrencies, the process of mining has two parts. Part one collects and validates pending transactions into a block. Part two calculates the hash of the block as the combination of the hashes of the transactions as well as a random value (a *nonce*) such that the hash of the block and the nonce matches a pattern–multiple leading 0 characters[15].

To make the game more fair for everyone, and to make sure that one really fast computer can't churn out answer after answer every second, the length of the prefix pattern varies. Maybe it's six leading 0 characters and maybe it's twelve if every device on the planet starts to mine Dogecoin. This is the *difficulty* of mining a block.

There's no reason you can't mine a block by hand, if you're quick at calculating things and extraordinarily lucky at guessing random numbers. It's just faster to use a computer when numbers get serious.

---

[15]Technically the hash must be less than a specific target value, but if it's helpful to think of it this way instead, it's still mostly true.

Tip #6   Interchange Your Wallet Keys

Every tip in this chapter so far has alluded to the fact that cryptography works best with large numbers, but have used smaller numbers to make it easier for you to count on your fingers, with a pad of paper, or in a spreadsheet. Big numbers are more difficult to manage without a computer–and big numbers are easier to get wrong.

Getting a big number wrong could be catastrophic. If you want to send your friend ten thousand Dogecoin to help them rebuild the roofs of their orphanage after a hurricane halfway around the world, you want to get their address right. If you're generating keys with an external process or tool such as `libdogecoin`, importing a private key or wallet backup, or doing something else to transfer your key or keys, wallet, you want to get the details exactly right.

Typing in or writing down or reading the long stream of numbers and letters that represents your private key is risky unless you double- and triple-check things. Alternately, you can use a different representation of your private key, designed to resist errors.

The Wallet Import Format

The *Wallet Import Format* or WIF encodes private keys in a shorter fashion and includes a checksum to help detect errors. You'll often see wallet backups and private key generators use this format. The Bitcoin wiki explains this well[16], but doesn't always explain why it works.

Here's how.

Imagine the hexadecimal representation of your private key is `0x123abc456def`. That's not a valid private key, but it's clearly a valid hex number and it's both long enough to demonstrate the principle and simple enough to keep in your head.

First, prepend the Dogecoin mainnet prefix of `1e`. This is a value hard-coded in Dogecoin source code, with variants for testnet and other networks. This gives `0x1e123abc456def`.

Now calculate the double-SHA256 hash (see Make a Hash of Fingerprints, pp. 1) of that value. Leave off `0x`, because that's a prefix that indicates this is a hexadecimal number. Convert the hex into a number, then run it through SHA-256 twice.

---

[16]See https://en.bitcoin.it/wiki/Wallet_import_format.

You'll get a large number back, but you only need the first eight digits (four bytes): 18d79603.

Append those eight digits to the end of the number with the prefix. You should get the result 0x1e123abc456def18d79603.

Hash Numbers, Not Text

Remember that a hexadecimal number represents a value. Check to see that your hashing algorithm properly executes on the *bytes* of the number itself, not the text value. The easiest way to check this is to swap the capitalization of any alphabetical characters in your hex string. 0xABC123 should hash to the same SHA-256 value as 0xabc123. If this doesn't happen, you're hashing the text value, not the number.

Finally, convert that new number (the original number with your network prefix and the checksum) to a Base-58 number (see Shorten Large Numbers, pp. 9). You should get 2gzKxzVzWxctr9.

This is the WIF-encoded version of your private key. Good wallet software can import from this format–and, importantly, tell you if the checksum is wrong. The checksum property also allows you to perform a decoding operation to verify all of the steps.

To validate the code, perform the reverse operation. Remove the four-byte checksum. Double-hash the remaining number. Compare the first four bytes of the resulting hash to the checksum. If everything matches, take off the network prefix and validate *that* against the expected network prefix. If everything matches, you have a valid WIF-encoded private key. Finally, attempt to import that key and see if it passes the network validity test (essentially "does this long number match the Dogecoin version of the elliptic curve equation?").

Nothing in any step here hides data or transforms it in an irreversible fashion. Given a private key in any step of this process, you can look at it and get that private key with pencil and paper (okay, maybe it's tricky to validate the double-SHA checksum by hand).

Understand the Risks

While not all wallets or Dogecoin clients support this approach, you could use the same technique to encode Dogecoin addresses. This shortens the addresses and adds checksums to verify their integrity.

Using WIF makes it easier to represent a private key, because it's less likely for you to get things wrong (it's shorter but not in an ambiguous way and it contains a checksum). However, it's still your private key.

If you're committing your private key to paper or to a file, you must still treat that artifact with all of the security you can bring to bear. Maybe you split the file into multiple pieces on multiple systems or write down the key in parts on multiple pieces of paper stored in separate locked safes.

Alternately, you can do away with keys altogether and use an HD wallet with a passphrase (see Use a HD Wallet, pp. 39).

Finally, be aware that you should only *decode* or *import* your keys on trusted systems with software you understand, validate, and control. If someone advertises an online WIF checksum validator, assume they're trying to steal your private keys. Stay away.

  # Tip #7   Create Asymmetric Keys

Hang around the cryptocurrency world long enough, and you'll hear people talk about public and private keys. Keys are cryptographic concepts used as metaphors (locks protect secrets, and locks have keys); in reality they're pairs of large numbers that increase your security. A key pair lets you use your private key to encrypt a message such that anyone with your public key can decrypt it. Similarly, anyone can encrypt a message to you with your public key and only you with your private key can decrypt it.

Every encryption operation is a one-way operation which requires one of the two keys and requires the other key to undo.

What magic is this?

Grab a calculator or a computer and find out.

RSA Encryption

An asymmetric encryption algorithm requires two keys. The keys are different; that's what makes the algorithm asymmetric. A symmetric approach uses the same key to encrypt a message and to decrypt the message, so there's no security other than keeping the single key secret. We can do better!

What would make asymmetric encryption effective? The encryption has to be a one-way mathematical operation. Given an encrypted message, you should not be able to decrypt it without the private key[17].

Assume that you want to send a secret message to someone, and that message is a number[18]. You want to use your private key, *d*, to encrypt the number. You want someone else to use your public key, *e* to decrypt the number. Is there a mathematical operation that can represent this activity without revealing the secret message or your private key?

One common asymmetric encryption algorithm is RSA. It's complicated, but one of the simpler versions. Fortunately, it's possible to understand the basics with little more than a calculator, a spreadsheet, or a sharp pencil and piece of graph paper.

RSA requires two prime numbers and modulus math (Roll Over Your Odometer, pp. 5). It uses these prime numbers to figure out the modulus number: the product of the two primes. For example, if you choose prime numbers of 5 and 17, your modulus is 85.

This modulus is used as half of both the public and private keys. The other halves of the keys require a little bit more math to figure out: exponents and reciprocals.

House of N, Powers of N

Think about exponents for a second. 10 to the power of 2 is 100, while 100 to the power of 1/2 is 100. Similarly, 2 to the power of 3 is 8, while 8 to the power of 1/3 is 2. This rule generalizes: *n* to the power of *y* equals *m*, while *m* to the power of *1/y* equals *n*. Work that out on paper or with a calculator to get a feel for it.

RSA uses this property as part of its encryption. Of course, if public and private keys were always obvious like this, it would be easy to turn a public key of 333 into 1/333 and reverse the encryption. That's where modulus math comes in.

Modulus of Powers of N

Think about how a clock or odometer works, especially how it throws away information. It throws away information, yes, but more importantly it *hides* the information it throws away. If the highest Donkey Kong level number tracked is 4, you can't tell if someone is on their first playthrough of level 4 or their 100th, unless you've watched their entire game.

---

[17] You could guess or spend trillions of dollars with supercomputers to guess, but even that would take a long time.

[18] It doesn't *have* to be a number in practice, but you can represent any message such as an email, an image, or a video as a number, so we'll continue to call it a number.

Throwing away data seems like it'd be at odds with encryption, at least if you want to get the right message out of the results, but the mathematicians Fermat and Euler found a solution.

Fermat's Little Theorem says that if you have two numbers, n and p, where p is prime and n is not a factor of p, then you can raise p to the power of n minus 1 and take the modulus of n. The result will be 1.

Here's where you might need to get out your pencil. Try this formula with n of 3 and p of 5. 5 to the power of 2 is 25, and 25 mod 3 is 1. Try it with 4 and 7. 7 to the power of 3 is 343, and 343 mod 4 is 1. Try it with larger numbers too if you like.

Euler liked this theorem and brought in a concept called *relatively prime* numbers. These are numbers that have no common factors, such as 2 and 3, 3 and 5, or 8 and 11. It's easy if the larger number is prime, of course. Euler's Theorem generalizes Fermat's Theorem to say that, given two numbers, x and n, where n is relatively prime to x and n is prime, then x to the power of n minus equals 1 mod n, just like in Fermat's Little Theorem.

Try it until it makes sense! Don't worry; there's no Vizzini voice about to say "Wait until I get started!" This is almost over. Try this yourself with a few interesting numbers.

Putting It All Together

That's enough math for now. What happens when you put it together?

The public key is the modulus number, N, and the public exponent, e (for *encrypt*). The private key is the private exponent d (for *decrypt*). The message is M.

To encrypt the message, raise the message to the power of the public exponent and take the modulus of the result: raise M to the power of e and modulus by N. Call the result C for *Crypted* message. To decrypt, raise the encrypted message to the power of the private exponent and take the modulus of the result: raise C to the power of d and modulus by N.

If you've chosen d and e and N appropriately, you will see the original message.

Where do the exponents and modulus come from? Start with a pair of prime numbers. For example, if the prime pair you start with is 5 and 17, the modulus is 85. Given that modulus, you can choose any number that is relatively prime to it for the public exponent e. In this case, 3 or 5 work well. Finally, to find d check for a number less than N where d times e modulus N equals 1[19]. Given a public key of 5, the private key is 13.

---

[19]Brute force? Really? It's not a math book. This approach works for small numbers.

Now for any message M that's an integer less than 13, encrypt it by raising it to e and taking the result modulus N. Decrypt that result by raising it to d and taking the result modulus N. You should get the original message back.

Try it with a few numbers. For example, encrypting 11 gets 61 and decrypting that gets 11.

You can also go the other way; encrypt with the *private* key. 11 encrypted with d gets 41 and decrypting that with e produces 11.

Limitations and Complications

For the math to work, the message to encode–the starting number, in this case–has to be less than the modulus, otherwise the math gets weird and you'll scratch your head at the results. In practice, the modulus is a really large number *and* any message larger than that number gets chunked into pieces and encrypted separately. For the purpose of this explanation, you don't have to worry about that–but in practice, the assumptions made here need some careful handling.

That's why it's important to leave the *implementation* of cryptography up to people who deeply understand the math and have thought through the important edge cases. It's easy to get things wrong.

By all means play with the math and write your own code and see what works and doesn't work–but don't use your own code for anything you really want to keep secure without doing lots of research and getting lots of peer review and understanding cryptography deeply.

What Can You Do With This?

While Dogecoin doesn't exactly do cryptography this way, it does cryptography similar to RSA, using a different algorithm called ECDSA. Once you understand the basics of asymmetric keys, you will know enough to know the shape of how Dogecoin and other cryptocurrency keys work.

Furthermore, understanding the limitations of RSA helps explain why ECDSA and other approaches are much better. As computers continue to get faster, or at least as computing power continues to get cheaper, older algorithms start to become less secure because computations that were expensive 20 or 30 years ago are now feasible on commodity hardware. 64 bits of security in 1978 are but a tickle to a modern phone in 2023.

Remember: the security of your Dogecoin depends on the security of your keys, both in terms of the strength of the cryptographic algorithm used as well as how well you keep your secrets secret.

Tip #8   Embrace Entropy

To keep a secret safe, you must limit the ability of anyone to access it, including guessing it. When you're dealing with the secret keys that keep your cryptocurrency under your control–and only your control–the better your secrets, the safer you are. Think of it this way: if Dogecoin only allowed four-digit passwords and those passwords had to use only the digits 0 - 9, there would be only 10,000 possible passwords.

Even a novice hacker could try all 10,000 possible passwords in a short time[20]

Established cryptocurrencies such as Dogecoin use much larger numbers. In theory, there are more possible numbers than there are atoms in the universe. Guessing any one specific number randomly will take a novice hacker quite a while[21].

Yet just because your secret can be a large number doesn't mean it's a secure number, especially depending on how you generated it. For example, in August 2023, researchers discovered a flaw[22] in secret generation in the Libbitcoin explorer bx. While that tool could generate appropriate secret passphrases for Bitcoin, Litecoin, Dogecoin, et cetera, it had a flaw that it generated far fewer possibilities than necessary–few enough that a clever hacker could reveal multiple secrets in hours.

The problem was a lack of entropy.

What is Entropy?

Entropy is the measure of randomness in a system[23]. The more random a system, cryptographers believe, the less predictable and more difficult it is to guess any item in that system.

It's important to distinguish between perceived entropy and actual entropy. In a truly random and cryptographically secure system, you might find a Dogecoin private key containing the string "12345" or "CAk3". That's not a bad thing; that's your brain pulling patterns out of apparent chaos. There are probably countless examples of these strings in public and private keys and countless other examples

---

[20]Definitely less than weeks. Minutes, probably.

[21]That means *never*, at least as we understand computers now.

[22]See https://cve.mitre.org/cgi-bin/cvename.cgi?name=CVE-2023-39910.

[23]In addition to the inspiration for some of Matt Bellamy's best songs.

where your brain might say "Hey, that looks familiar!" That doesn't mean there's not enough entropy.

Why is entropy useful though?

Ten Pounds of Potatoes in a Five Pound Bag

Let's try an experiment. Find a SHA-256 hash generator (see Shorten Large Numbers, pp. 9). This algorithm produces a 256-bit hash string on any input you provide. That's a large number. That should be difficult to guess, right?

Try an input of 1, then 2, and so forth, up to 9. Look at the outputs. They all look pretty random, don't they? Yet imagine using any of those as inputs to your private keys. If someone knew you started with a single-digit input, how long would it take someone to generate all of the potential keys you might have used? Not any longer than it took you to generate those hashes.

That's why entropy is so essential: you want true random input with enough of a range of possible values that the number of things an attacker would have to search to find anything useful is so large that it's effectively impossible.

Computers can't generate *truly* random numbers. They have to use an algorithm with plenty of input to generate something that's sufficiently random it can be useful for cryptographic applications. The problem with the Libbitcoin Explorer is that it limited the range of entropy it used as input to only 32-bits of data. That's better than the digits 1 through 9, but that's a small enough domain that a fast computer can test all possibilities quickly.

The ironic thing about entropy is that it's more useful when you have enough you can *throw away data*. If you're trying to get a 256-bit random number, having 256 bits of data is good. Having 257 is better. Having 512 is better still, and so on.

Think of this like modulus math (see Roll Over Your Odometer, pp. 5 and Create Asymmetric Keys, pp. 17). Reverse-engineering a modulus math problem is, as we understand it now, intractable. You can't know how many times someone has gone around the range of the modulus.

What Can You Do With This?

What do these heady math concepts mean for friendly/funny dog money? They mean everything, at least when it comes to your security. If you didn't understand entropy and its value, reading through the security advisory linked earlier would be difficult. Similarly, if you didn't understand that a hashing algorithm can only do so much with what it has, you might be tempted to use "sw0rdfish" as your password[24] and call it a day.

Even if you don't delve deeply into understanding cryptography and math, now you know that a 256-bit private key needs at least 256 bits of entropy to protect your secrets. Use that knowledge as you explore tools and techniques available to you. Reliable and trustworthy software will go out of its way to explain the proper handling of entropy.

  # Tip #9   Identify a Scam

Some of the most useful features of a cryptocurrency such as Dogecoin, including the opportunity for anonymity, lack of a single centralized authority, an immutable transaction ledger, and programmable money have their risks. The same is true of cash, credit cards, and other payment instruments.

Yet for all of the promise that a decentralized cryptocurrency means that you don't need anyone's permission to make a transaction, there's also no person to appeal to if something goes wrong. There's only math and network consensus, and they only care about accuracy, not ethics. If you have a working motto, it should be "don't trust; verify".

In short: it's your responsibility to make sure that you avoid scams and avoid scamming people (see Pledge to Do Only Good, pp. 230). Here are some ways to help you do that.

What Makes a Scam?

Every scam shares one common element: someone asks you to give up control of your funds. They promise one thing or another, but every time someone wants to get their hands on your money, currency, coins, whatever–they eventually need you to hand it over. This can take many forms, all of them promising you something amazing in the future in exchange for something valuable now!

Buy Low, Sell High

A "fear of missing out" scam, or FOMO, puts pressure on you to buy something now because the price will rise in the future and you will lose your chance at for great returns. You'll kick yourself in the future, when everyone else is driving a Lamborghini and sipping muddled blueberry lemonade by their dollar-sign shaped infinity pools. "Have fun being poor", people say, as they put more pressure on

---

[24]Please, never do this!

23

you to think about the regret you will eventually have for *not* giving them your money.

It's natural to think about all of the amazing things you can do if everything goes right. It's easy to get caught up in the excitement. That's why lottery tickets sell. Even though your odds of winning are always very, very low, the idea of winning big is so exciting that a couple of dollars seems like a small price to pay. As well, people *do* win: just not very many.

How do you resist this? Demand answers to questions such as "why will the price go up" and "what is the evidence that the price will go up" and "who benefits if the price goes up". If the answers all hand-wave away your concerns, increase your skepticism. "Demand will increase" is not an answer, unless there's some reason *why* demand will increase, and "people will see the price continue to go up" is *not* a reason why demand will increase–it's evidence that the scam will continue.

Give Me Your Money, I'll Give You More Money

Another scam locks up your money under someone else's control with the promise of incredibly high returns. "You'll earn 60% returns in a year!" they say. This sounds amazing, even as which every scam detector in a twelve-block radius goes off. If you can earn 60% returns in a year, why would you need to ask anyone else for money? Why would you need to control someone else's money to earn those returns? (Even a credit card at a punishing rate will only charge you 30% APY or so, so your effective return before taxes will still be extraordinary.)

You can often see this scam in terms of "staking" coins, where you promise to give up custody and not sell in exchange for payment down the line. What they aren't telling you is that *they* will sell or lend your coins, and they're going to pay you with the proceeds they get from other people.

How do you resist this? Demand answers to questions such as "where do the profits come from". If you stake 100 Dogecoin on January 1 and someone promises to pay you back plus 60 Dogecoin on December 31, they ought to be able to explain exactly where those 60 Dogecoin come from (as well as any other profits they make on the transaction, because you know they're not giving you all the profits).

Clone, Pump, and Dump

Yet another scam trades on the history, goodwill, and name recognition of one thing while promoting an alternative that isn't the same thing at all. In fact, the alternative isn't the same thing so much that it's entirely under the control of someone or someones who stand to profit from the confusion.

Imagine, for example, that Pupper Soda were a new soft drink brand, and it had a passionate, growing following of people who appreciated its bold, fresh taste, low-calorie presence, and need for very little sugar. More and more restaurants, taco stands, and vending machines offered Pupper Soda, and its creators reaped the financial rewards.

Someone approaches you on the street as you're enjoying a refreshing beverage and said "I bet you wish you'd invested in Pupper Soda." You nod, wary. "Well, they're coming out with a new product, Good Pupper Soda, and it's going to be even bigger." Your eyes narrow, and you avoid the temptation to put your hand on your wallet to keep it close to you. "You should buy into Good Pupper Soda while you can, because it's going to be huge."

You close your eyes, because you know he's about to wave a contract in front of your face and demand you hand over whatever's in your wallet, without acknowledging that he made a copy of the Pupper Soda vending machine, changed the name, and is filling up the cans with tap water from the hose on his garage[25].

You're going to get a Fear of Missing Out pitch shortly, and you're going to hear "Have fun being poor" and, unless you walk away quickly, you'll get a lot of hand-waving explanations about why Pupper Soda is going to be huge. Your conversational assailant is in the Pump phase of Pump and Dump, and you're the target.

How do you resist this? Demand answers to questions such as "why do you need additional investors" and "who owns what you're selling" and "why this specific knockoff, instead of any of a thousand others". If you're interested in getting deep into technical details, ask "What implementation makes this different from the original", because it's easy to shave the serial numbers off of someone else's work and slap a fresh coat of paint on it and pass it off as something original[26]. Finally, ask "Who controls it" and *verify* the answers.

In the cryptocurrency world, if someone has made a new coin (or, even easier, a new token) and spent more time making a logo and website than making something different in terms of protocol, consensus, or implementation, you're probably probably looking at the kind of scam where they'll sell you their supply and then disappear. They get your money and you get a bag full of nothing.

Not Your Wallet, Not Your Coins

Another scam or scam-adjacent risk is when someone requires you to hand over control to your wallet, keys, and/or coins. This could be anything from holding

---

[25] Admittedly, this analogy is getting away from the author.

[26] See, for example, every AI trained on copyrighted data without permission.

funds in a tipbot on Discord or another social media platform, holding funds in an exchange, or letting someone else see your private keys for any purpose other than backups[27].

Anyone who holds your keys can do whatever they want. The network, the protocol, and the network consensus protocol cannot distinguish between what you intended and what the math allows. **The network will not protect you.** The network does not distinguish between what you consider a "right" or "wrong" use of your keys. Either the transaction validates or it doesn't.

How do you resist this? Give no one your keys and give no one control of your wallet. If you can't avoid this, reduce access temporarily by transferring your coins to a wallet completely under your control, practicing safe key management discipline, and distrusting anyone who tells you that you must trust them.

What Can You Do With This?

This isn't an exhaustive list of all possible scams. There are many more, with variants appearing all the time. It's worth repeating: all scams depend on you giving up control somewhere, somehow to someone else.

You work hard for what you have (see Make Money with Dogecoin, pp. 252), and it's important to protect your interests. Check and double-check what you intend to do. Finally, remember that you can only trust the people and things you can verify. Anyone operating with clear intentions should make it easy for you to verify what they say. Anyone who hides their intentions between "trust me" or "what are you afraid of, success" is waving a red flag.

  Tip #10   Roll the Dice

Randomness is essential to cryptography, and randomness is essential to the world around us. Think about flipping a coin, spinning a game wheel, or rolling dice. Over time, you should get a roughly even distribution of results (half heads or tails, 10% of the time on each of ten pie slice-shaped wedges in the wheel, or one-sixth of the time on each of six sides of a die)–assuming the randomness is truly random and there are no external factors in play, such as a weighted die.

Randomness when generating passphrases and private keys helps you have your

---

[27] If you've distributed fragments of your private keys to your brother, your attorney, and your childhood friend in case anything happens to you–and you do the same for them–that's a different story, though one with its own risks.

own individual secrets (so no one else can spend your coins), and randomness when mining blocks helps keep the network secure, so no one individual or team can control all mining (see Host a Puzzle Race, pp. 11).

How does this randomness play out in practice? How can you tell the difference between fake randomness (pseudorandomness) and real randomness? It's all about entropy (see Embrace Entropy, pp. 21)–and it's easy to see how this works in practice with the right example.

One Page RPGs

A game designer named Oliver Darkshire has a hobby of creating one page RPGs[28], where you need a single six-sided die, some pennies or other counters, and a few minutes to play a game. These five-minute workday breaks each tell a fun little story and are a treat to play. While the rules are simple enough to explain in a paragraph, the story that emerges through the gameplay is slightly different each time.

That sounds like randomness, doesn't it? It's not just the randomness of the die you roll, but the choices you make. If everyone made the same choices and the rules produced the same results every time, it wouldn't be a game. It would be Candyland, and where's the fun in that?

One Screen RPGs

Suppose you don't have a die handy, but you do have a laptop with Ruby installed. You could write a little program to roll a series of dice for you to play one of Oliver's RPGs. It might look like this:

```ruby
require 'games_dice'
require 'tty-prompt'

def main(random_seed = '0x01')
    srand(random_seed.to_i(16))

    items = %w(1 2 3 4 5 6 quit))

    while true do
        prompt = TTY::Prompt.new
        input = prompt.select(
          'Choose number of d6 to roll', items, per_page: 7
        )
        break if input == 'quit'
```

---

[28] See https://www.patreon.com/deathbybadger and https://x.com/deathbybadger.

```
        input.to_i.times do
            dice = GamesDice.create '1d6'
            puts dice.roll
        end
    end
end

if $PROGRAM_NAME == __FILE__
    main(*ARGV)
end
```

Use either gem install or bundler to install the games_dice and tty-prompt libraries, then run the program. You'll see a text-based menu asking how many six-sided dice to roll:

```
Choose number of d6 to roll
(Press ↑/↓/←/→ arrow to move and Enter to select)
 > 1
   2
   3
   4
   5
   6
   quit
```

Select the quit option to exit the program. Otherwise, use the arrow keys to select the number of dice to roll and use them as the rolls for your game.

If you quit the program and start over, you'll notice something interesting: the output is the same every time. Choose 6. On the author's machine, he sees rolls of 6, 4, 5, 1, 2, and 4. Start the program again and choose two rolls of 3. Your author gets results of 6, 4, 5 and 1, 2, 4. Those are the same rolls in the same order.

This seems not ideal, but it's an inherent property of the program as written:

```
def main(random_seed = '0x01')
    srand(random_seed.to_i(16))
```

By default, Ruby doesn't use a truly random number generator. It uses a pseudo-random number generator that, when provided a starting number, produces a series of outputs that seem random and unpredictable. Yet if you know the starting point, you can always get the same results. That's what srand does; it fixes a starting point for Ruby. If you don't provide one, the program will always use the same value.

Removing the srand line will produce different results each time you run the program—but even that isn't truly random. It's just a different starting point that produces a different series of outputs. For a game, this is probably fine, but for cryptography, it's not.

Fortunately, we have a better source of entropy available.

Numbers Used Once

When miners mine blocks, they have to solve puzzles, and the answer to any block's puzzle is a *nonce*, a shortening of the phrase "number used once". A good nonce should be unique, as obviously as possible random, and nothing someone could predict trivially by looking at a block beforehand. The entire goal is that computers will have to test a lot of difference nonces before they find one that will solve the puzzle.

Given all of that, could a nonce be used as a source of entropy? It's a number, it's unpredictable, and it's something no one can predict until it appears in a mined block. You could write a little launcher for the dice game like:

```
#!/bin/bash

hash=$(dogecoin-cli getbestblockhash)
block=$(dogecoin-cli getblock $hash)
nonce=$(echo $block | jq -r .nonce)

if [ "$nonce" -eq "0" ]; then
    nonce=$(echo $block | jq -r .auxpow.parentblock)
fi

bundler exec ruby ruby_dice.rb $nonce
```

Don't worry if you haven't read ahead yet (see Command the Core, pp. 42, to start); you don't have to understand the details of how this works to understand the concept.

┌─ Merged Mining ───┐

If you've read ahead and played with the code yourself, you might notice that `nonce` is often 0. This isn't a bug; it's a feature of something called merged mining, where miners can mine multiple blockchains such as Litecoin and Dogecoin together and use proof of mining for one chain to prove work for another. In that case, the random value used for the dice roller is the hash of the parent block–still a good random seed.

└───┘

What Can You Do With This?

Rolling dice takes only a couple of lines of code, even without the nice Ruby libraries included here. If you're looking for an interesting weekend project, try turning one of the one page RPGs into a full-fledged game, using the skeleton program already provided–though if you do, don't hard-code the random seed; call srand() only if someone has provided a non-default value.

Admittedly, using a blockchain value mined every 60 seconds or so is a lot of work to get a random number for something as frivolous as a dice rolling program. In practice, Ruby's (or Perl's or Python's or Node's or...) built-in pseudo-random number generator is sufficiently random that you'll have a fun game.

Even so, it's important to recognize the limitations of pseudo-randomness. Where's the fun in a game where you're trying to get rid of a Mastodon but the outcome is always predetermined? It's not a game; it's a story. Similarly, the fun of generating a passphrase but realizing that someone else has already guessed it because you used a system with predictable fake randomness is infuriating and disheartening, not fun.

By looking for places where true randomness is essential–and where it's not–you can identify where you need to be cautious about what produces that randomness.

CHAPTER *2*

Running Your Own Node

Dogecoin gives you as much or as little control and custody of your funds as you like. You can delegate management of everything to an exchange, for example, or you can run your own node and have complete control over your funds and even how the network as a whole operators. While taking matters into your own hands may seem complex, risky, and daunting, it can be both easier and safer than you think, and it gives you a lot more options.

This chapter demonstrates interesting things you can do when you download the Dogecoin Core software and use it to run your own node.

  Tip #11   Understand Core Programs

A released version of the Dogecoin Core software unpacks on your computer with several files in the *bin/* directory. Any guide which explains how to run a node (such as Run a Node, pp. 34) will mention one of two files in that directory, but it may not explain the other files, why they exist, and what you can do with them.

Understanding what you have when you have the Core software installed will give you important options.

Common Binaries

You're likely to use one of two binary files all the time, either dogecoind or dogecoin-qt. Why choose one over the other? That depends what you want to do.

Not all release downloads contain all of these programs. If you downloaded a Dogecoin Core compiled for a Raspberry Pi, for example, you may not have dogecoin-qt available. This generally happens because the libraries needed to run the GUI aren't available for that particular platform. This may change in the future; check the release notes for the version you're using.

If You Want to Manage a Wallet

You don't *need* to run a node to manage a wallet, and your wallet doesn't need to be connected to a running node to work. However, running a node with a wallet connected (and carefully secured) can give you priceless insight into your transactions and transaction history.

The `dogecoin-qt` binary is the Dogecoin Core GUI node. It supports wallet features and lets you manage wallet addresses, track sent and received transactions, and perform all other node activities. Almost everything you can do with a node is available through the GUI (and sometimes the *dogecoin.conf* configuration file).

If you're comfortable running desktop software, this is a good choice. You don't have to run it with a wallet enabled, but most people who run it do so to use its wallet.

The `dogecoind` binary is the non-GUI version of the core. It supports wallet operations as well, but this program's primary interface to any Core behavior is through the command line, via RPC calls (see Command the Core, pp. 42), et cetera. If you're running a node *not* connected to a wallet, this may be a better choice.

If you're running a node on a remote server, such as a cloud server, this may be your only choice.

Both `dogecoind` and `dogecoin-qt` are long-running, server-style processes. Depending how you have them configured, they may take several minutes to start up before they give you the ability to explore your wallet and the network. This is because they need to connect to other nodes, validate data they have, download new blocks since they last ran, and perform other housekeeping exercises.

If You Want to Get Node and Network Data

If you're running a node and want to get information out of it—especially without using the GUI—use the `dogecoin-cli` binary. It can connect to a running node (either `dogecoind` or `dogecoin-qt`) and issue commands. That node can be running on the same machine or on a remote machine, whether on the same network or available from the Internet.

By itself, the program won't do much beyond respond to `--help` commands. To get real data from a node, you *must* configure both the node and your network securely so that `dogecoin-cli` can make RPC connections (see Authenticate RPC Securely, pp. 73).

There's no GUI for this program. It's useful from the command line. You can also invoke it from a programming language, though you may find it more useful to make RPC calls into the core directly.

If You Want to Work with Transactions

You can create and sign transactions with dogecoin-cli and dogecoin-qt, but they both require a running node. To manipulate transactions *without* having to connect to a node, use the dogecoin-tx binary. While you *do* have to connect to the network somehow to broadcast and receive transactions and blocks, you can work with a wallet and transactions offline and only connect to the network when you've made all of the changes necessary.

In other words, if you decide that you need the security of keeping your wallet information on an offline device, you have the option of running dogecoin-tx on a machine not connected to the Internet or any network, then copying the transaction you want to make to a different machine with no access to your wallet and using that second machine to send your transaction to the network.

If You Want to Debug Something Weird

Finally, the test_dogecoin binary runs a series of tests that the Core developers use to increase their confidence that the software works as expected. All tests must pass before a release (tests must also pass before the Core developers accept a change to the code *before* a release as well), but the variety of computers and hardware and software and configurations mean that there's a chance you might experience a bug the developers didn't anticipate or experience.

If you find something behaving very much not as you expected, running this program may provide you and developers with useful debugging information. However, be aware that any bug you find will be a surprise, so back up your wallet (or, perhaps, move your wallet file *off* of your computer) before running this program.

When you have the useful information, open an issue (see Open an Issue, pp. 118) and see what you can do to help ensure no one else encounters the same problem.

What Can You Do With This?

Most Core users will probably use dogecoin-qt and be happy with it, and that's perfectly fine! You can get a lot done with that program, and you always have the option of modifying your configuration or changing options to suit your needs.

When you want or need to do other things—a lot of them documented in this book—then having other programs and options available is important. Even though your first experience with the Core might be running a node with a wallet attached, you have plenty of options. Those options let you change your experience based on your desire for privacy and security, any advanced uses you have, or even the time and effort and resources you can spend maintaining a running node.

Tip #12   Run a Node

A subtle but essential truths of Dogecoin is that Dogecoin is what we all agree it is. While developers have some ability to add features, fix bugs, and release software for other people to examine, run, modify, and redistribute, it's the *network*–the collective behavior of everyone who participates–that decides what actually happens. By design, everyone who runs a node gets a say in which behaviors are valid or invalid.

For example, the Dogecoin Core 1.14.5 release lowered the default fee per transaction from 1 Doge per transaction to 0.01 Doge per kilobyte of transaction size. You can now take advantage of these fee changes because enough nodes have adopted the new version that the network supports these lower costs.

The power of the network–what we call *consensus*–is in the hands of people who run nodes. That could be you.

What You Need

What do you need to run a node? A computer. Some memory. An Internet connection. Persistent storage. The willingness to do some research and keep things up to date. Patience. A kind heart.

In specific, you need a relatively modern computer (a decent desktop or laptop machine manufactured in the past 10 years will work). As of this writing in April 2023, a full node requires over 60 GB of free hard drive space. The size of that requirement will grow in the future. 4 GB of RAM will help. Windows, Linux, and Mac OS X are all supported. Other operating systems or architecture combinations may require more work on your part.

> **Are you a node-half full person?**
>
> What's the difference between a full node and a partial node? If your node stores the entire blockchain history from the first block mined until today, you have a full node. Otherwise you have a partial node.
> While running a partial node saves disk space and provides some benefit to the network by validating and transmitting transactions, full nodes are essential to verify the validity of *every* transaction and to help other full nodes come online. If you can spare the space, running a full node is a great contribution.

Find the Right Software

As of this writing, both `dogecoin.com`[1] and the Core GitHub repository[2] are reputable sources which announce new Dogecoin Core releases and provide download links to Dogecoin Core releases. Depending on when you read this, one or both of those URLs may have changed–probably not, but it's possible–so do some research to figure out what's reputable.

From a verifiable source, download the distribution that best matches your operating system. For example, if you're running Windows on Intel or AMD hardware (not ARM), look for a `win64` bundle. For Mac, look for the `osx` bundle. For Linux, look for the `linux` bundle that best matches your processor bundle. Then verify the download (see Verify Core Releases, pp. 101). If something looks suspicious or seems off, stop! Do some research. Figure out what went weird, then decide if and how you want to try again.

After you unpack the software (Windows has an installer or a zip file, Mac OS X has a DMG file, Linux users get tarballs), you'll end up with a couple of alternatives. Do you want to run a GUI or a background process?

This book assumes you'll run the GUI but have access to the other files: a command-line interface called `dogecoin-cli` and the background process `dogecoind` (see Understand Core Programs, pp. 31).

> **Are There More Details?**
>
> If these rules seem like they assume too much knowledge you don't have yet, or if things look very different in the future, look for a Dogepedia entry called "Operate a Dogecoin Node"[a]. That guide goes into more detail and, because it's a website, can be updated when things change more quickly than a printed book can.
>
> ---
>
> [a]Currently at https://dogecoin.com/dogepedia/how-tos/operating-a-node/.

Configure Your Node

When you first start your node, you may see a prompt asking you where to store configuration information, logs, and persistent data such as your wallet and the blockchain. Take note of this.

---

[1] See https://dogecoin.com/.

[2] See https://github.com/dogecoin/dogecoin.

- On Unix-like systems (Linux, BSD, etc), the default is *$HOME/.dogecoin/*

- On Windows, the default is *%APPDATA%\Dogecoin*

- On Mac OS, the default is *$HOME/Library/Application Support/Dogecoin*

You can change these if you like, but you run the risk of confusion. If, however, you need the hard drive space elsewhere, changing this directory can be useful. The choice is yours.

After you have your node up and running, you'll have to wait a while to download the entire blockchain and store it on your disk. This can be a good time to start a new hobby or polish your skills with an existing one, such as reading a book, knitting a warm pair of socks, or baking a delicious pear galette. The secret to the latter is ginger and brown sugar.

You can also spend some time configuring your node.

For example, in the GUI, click on Settings then Options then Network. Click the box for "Allow incoming connections". Without this, your node will only receive data, never transmit. That can be useful to you, but it's not useful to the network. Be aware that you may have to do some work on your own network setup to finish the task, however: your node needs to be reachable from the public Internet on port 22556, and your machine needs a reliable IP address from your DHCP server.

Sometimes clicking the "Allow UPnP" box in the network tab in the Core GUI will fix this. Sometimes it won't. If this seems like Star Trek-style technobabble to you, that's fine. It's okay to stop at this point for a while; you can still *use* a node running on your own even if you don't or can't allow incoming connections. Ask a friend, do some research, learn the implications of what this means, and then decide if you want to continue.

Skim some of the other configuration options. They all have their uses. Other tips in this book will cover some of the most important. Do be aware that you will have to restart your node to take advantage of them, so pick a time when that's least disruptive.

Do make note of your configuration and data storage directory however. You'll use this a lot throughout the rest of the book.

Alternatives

You don't need a desktop computer or laptop. You could run a VM in the cloud. Various service providers such as Amazon AWS, Google GCP, Microsoft Azure, and Oracle Cloud offer free introductory packages as well as modestly-priced services, depending on your definition of "modest". You'll need some degree of system administration and automation skills to set up and run a node with these services, especially if you're sensitive to price caps and spending limits.

At the risk of referring to something that seems exciting (at the time of writing) but hasn't shipped a tangible result yet (at the time of writing), a do-it-yourself hardware project under the umbrella of very dot engineer[3] is attempting to assemble a known-working combination of hardware and software to run a full node with small form-factor computers, solar power, and goat-resistant networking[4].

If you're good with a soldering iron or flashing bootloaders or comparing spec sheets for all-in-one processor boards, check that project for more details.

Understand the Risks

Running a node has a few risks and costs.

First, it's a commitment of time and resources. Depending on the speed of your hardware and network connection, it could take a few days to download all of the blockchain. Depending on how well-connected your node is to other nodes, you could send dozens of gigabytes of data over your connection every month. Depending on your hardware, you could see a measurable increase in your power bill. Before you start running a node, consider what you can commit to now and measure carefully the effects (see Know Your Limits, pp. 44); take care of yourself first before the network.

Second, remember that running a network service like a Dogecoin Core node means you'll receive traffic from all over the Internet and you'll send traffic all over the Internet. While the Internet is full of wonderful, selfless people like everyone running Core nodes, it has its share of malicious people (some deliberately, some unknowingly), so keep your security practices up to date. Invest time and effort into a good firewall, monitor for suspicious activity, and don't perform any actions you haven't researched and vetted.

Third, keep your node secure (Authenticate RPC Securely, pp. 73, for example) and your wallet safe (Work Without a Wallet, pp. 151, for example). You can do a lot of interesting things with a Core node, but many of those interesting things are a lot more interesting when only you can do them. Otherwise they're more scary than interesting.

Finally, remember to keep your node up to date. Core releases add new features, of course, but they also fix bugs, add more configuration options, and improve security. One important thread in development philosophy is to give users–people

---

[3] Seriously, see https://very.engineer/.

[4] Possibly other livestock too. Perhaps even birds.

who run nodes like you are considering–the ability to shape the behavior of Dogecoin and its network. To do this responsibly, you need to understand what and how you can contribute to your vision of Dogecoin.

  # Tip #13   Set Your Node Comment

Look at the network peers tab in your core wallet (Help -> Debug -> Peers, in the Qt GUI as of the current 1.14.9 release). You'll see a list of other nodes you've connected to, along with like their IP addresses (IPv4 or IPv6), the amount of data sent and received, and more.

That additional data includes a node version string. By default[5], every Dogecoin Core reports itself as `Shibetoshi/version`. This applies whether you're running `dogecoind` as a headless network service or `dogecoin-qt` as a GUI application or some other way of running the Core that we haven't invented yet.

The Core allows you to *append* text to this version string to customize your node, amuse yourself, and, potentially, brighten someone's day through a feature called `uacomment`.

Setting `uacomment`

To customize this feature, you need to edit your *dogecoin.conf* configuration file and add an entry:

```
uacomment=My Cool Node
```

Of course, you can and should pick something more interesting than that. As it currently stands, you have 256 characters to play with here[6].

Save this file, then launch or restart your node. If you're running the Qt GUI, look at the debug information window under "User Agent" (Help -> Debug -> Information). The text you see there will be broadcast to other nodes whenever they connect to your node, so this is your chance to make a mark and do something meaningful.

---

[5] In the Dogecoin Core source code, look for a value called `CLIENT_NODE`.

[6] This is defined as `MAX_SUBVERSION_LENGTH` in the source code. Check the source for the version you're running to see if this has changed.

Setting a *Secure* UA Comment

Of course, this is your node broadcasting something potentially specific to you, so remember that you're broadcasting it to an entire Internet full of people, some of whom may not have your best interests at heart.

When you're choosing what to say, keep at least two important points in mind:

- Be Kind

- Be Safe

If you're thinking of setting a comment that might hurt someone's feelings or cause drama like "Patrick is a curmudgeon![7]", think twice. There are human beings on the other side of the Internet, and our funny dog money only works if we all work together and cooperate.

If you're thinking of setting a comment that might reveal personal or private information about you or someone else, like "Proudly hosted at 1 Chome-2 Dōgenzaka, Shibuya City", think twice, then three and four times. Personal information can make you or anyone else a target. A visit in person or electronically from someone who doesn't have everyone's best interest at heart goes against the spirit of fun and collaboration.

What makes a *good* comment? If you follow those three rules, then anything goes.

You could use a dopey joke, like "Why does the Swedish Chef love Dogecoin? Because it goes bork bork bork!"

You could advertise your Doge-related project: "Check out IfDogeThenWow!"

You could brag a little about your uptime, like "Proudly serving the network since 2013."

The sky is the limit, as long as you're smart about it.

  Tip #14   Use a HD Wallet

The safety of your wallet is the safety of your Dogecoin. The privacy of your addresses is your privacy. If you make transactions securely, keeping your wallet safe and creating new addresses for new transactions, you'll reduce your risk and improve your privacy and security.

---

[7]He'll probably agree, so this is the worst thing your author wanted to put in print.

Security and convenience aren't always tied together, but Dogecoin Core tries to help you out by generating a pool of unused addresses whenever it creates a new wallet. You can generate new addresses anytime, of course. The *way* the Core generates address is important; one approach is much more flexible and secure than the other.

You should use it.

Hierarchical Deterministic Wallets

In the olden days when Bitcoin was new, the first way a Core generated addresses was to generate a list of random addresses. By default, that used to be 100 at a time so that you could have new addresses ready without waiting to make one more–to reduce the temptation to reuse an address.

Then came BIP-32[8], a proposal to use a secret *seed* value to generate millions of private and public keys and addresses both to reduce the risk of running out of addresses and to reduce the complexity of managing wallets. You should understand two features of this proposal.

First, using a well-defined derivation approach from single starting point creates a *hierarchical* mechanism for generating millions of addresses. The combination of the starting point and a path along the list of all addresses makes it unambiguous what you're going to get.

Second, the mechanism used to derive addresses always gives you the same output with the same inputs. That's the *deterministic* part of the HD wallet. With the starting point and derivation path, you can re-derive the public and private keys and corresponding addresses.

All of the implications are important, but the most important part right now is that a wallet like this can come from a single *seed phrase*. If you have that phrase (memorized, stored securely in multiple places), you have your wallet *even if you never connect it to a computer again*.

How do you get an HD wallet?

First, check to see if you have one already. Launch Dogecoin QT and look in the bottom right corner. You should see a little image that reads "HD", as seen in *dogecoin-core-hd-wallet*. If you don't see this, make sure you're running a recent enough version of the Core (at this writing, at least 1.14.9).

---

[8] See https://github.com/bitcoin/bips/blob/master/bip-0032.mediawiki.

Figure 2.1: Dogecoin Core with an HD Wallet

Transferring All Funds to an HD Wallet

The long way to do things is to create an entirely new wallet, ensure it's an HD wallet, and transfer all funds from your old addresses to one or more new addresses. Sometimes this isn't feasible, and doing this may involve swapping wallets, using multiple profiles/computers, and paying transaction fees.

Upgrading Your Wallet with `upgradewallet`

If you *are* running a recent version of the Core but have an old wallet, you can upgrade your wallet. First, *back up your wallet*. Then close the Core and launch it again from the command line, adding the `upgradewallet` flag:

```
$ dogecoin-qt -upgradewallet
```

When the command finishes, use your Core as you always do; it's safe to keep running if you started it with this option.

While it's *safe* to use the upgrade command multiple times, there's no reason to do so if you've already upgraded to an HD wallet, so avoid unnecessary work and use the command only once. This command is *idempotent*. In developer-language, this means "you can use it more than once, and it won't do anything bad if you do". Of course, it *will* rewrite your wallet needlessly, so don't add this option to your configuration file.

Understand the Risks

Any time you ask the Core to make changes to your wallet, there's a tiny risk something could go wrong. Before you modify your wallet or create a new wallet, think about what you're doing, make and test backups, and plan for potential recovery.

Know where your backups are going, how long you need to keep them around. After you've made the change, check that all of your transactions and addresses are behaving as you expected. When you're comfortable that the changes are good, consider if you need to remove backups; there's nothing scarier than realizing that that USB key you grabbed just to be on the safe side fell out of your backpack at the airport[9].

  # Tip #15   Command the Core

The first time you launched the Dogecoin Core[10], you saw a friendly dog graphic and some text and that's about it. There are menus and options and buttons to send or receive transactions.

You may think to yourself "I know there's a lot going on behind the scenes, but is this it? A friendly dog image, a couple of buttons, and a lot of math I really ought to go back and read in more detail (see Understanding Cryptography, pp. 1)?"

Click the Help menu, then the Debug window item, and finally the Console tab. You will then see something like The Dogecoin Core Debug Console, pp. 42.

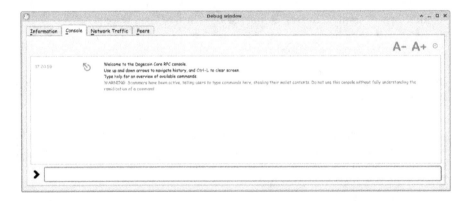

Figure 2.2: The Dogecoin Core Debug Console

---

[9]Your author has never done *this*. It was a cool hat.

[10]Maybe that's today. If so, congratulations!

What's the Debug Console?

What's going on here? This GUI says it's risky, but here it is, early in this book, and you're looking at it, wondering what's going on.

Behind the scenes, the Dogecoin Core does a lot of operations that you might expect, including:

- Managing your wallet

- Receiving incoming transactions

- Generating keys

- Keeping a list of addresses

- Organizing transactions into blocks

All of those operations necessary to keep the Dogecoin blockchain running and your data easily accessible. You're looking at a console where you can ask the Core to perform some of those operations and make queries on your behalf as you need them.

For example, if you want to know the height of the most recent block your node has processed, type getblockcount in the console window. You'll notice the input area text will start to suggest commands as you type more and more characters. When you finish, press the Enter key and you should see a response like:

```
18:28:57     getblockcount

18:28:57     4649991
```

What just happened? You sent something called an RPC[11] command to the Core and it responded. How did you know what to type? How did you know what response you expected?

Type help getblockcount to see the documentation for this specific command. Type help to get a list of all commands you can use in this window. When you're finished, close the window. In the meantime, your node will continue running and processing transactions, so if your node is connected to the network and you wait a couple of minutes and type getblockcount again, you'll get a higher number.

---

[11]RPC stands for "Remote Procedure Call", which means "ask a running program to do something". In this case, it returns a number.

Understand the Risks

Several tips throughout the rest of this book discuss RPC commands because they're useful in the right circumstances. While a lot of these tips assume you'll use the Dogecoin CLI client or an external program to execute the commands, you can use the console window too.

With that said, any command you type in here can cause your node to do things. If you have a wallet attached to your node and if you have Dogecoin in your wallet, some commands you type here could do things with that wallet and those coins. *Never run a command you don't understand* and *never run a command you don't trust*.

Even all the commands in this book you can and should verify. Use the `help` command to see the documentation on what the command does and think about the risks. This is especially true if you're reading this book and using a Dogecoin Core version other than something in the 1.14.x series or even 1.21.x.

  Tip #16   Know Your Limits

You've decided to do something selfless, maybe because you believe in giving back, maybe to bring about a decentralized consensus currency, maybe to learn something, and/or maybe other reasons. You've decided to run your own node (Run a Node, pp. 34). You've thought about the effort and resources you have available, and you know you can do it *if* you can manage the commitment.

Good news! While running a full node with no constraints is the best possible way to run a node, the network gets stronger with more nodes even if they're not all 100% always full and available. You can change how you want to contribute with a few tweaks and configurations.

Limit Node Disk Usage

A full Dogecoin Core node–a node that stores the entire blockchain from the first block until today–takes up some 100GB of disk space (as of November 2023). That number continues to grow, transaction by transaction and block by block. If you don't currently have enough disk space to devote to Dogecoin (or where it might grow before you can add more storage space), you can ask the Core to limit its stored history.

The `prune` command-line argument (or configuration file option) allows you to choose between three values. The default, 0, tells the Core to store every block, keeping everything, with no limits. The value 1 tells the core to allow you to issue prune commands on your own, giving you control. The third value, anything over about 3000 or so[12] tells the Core to keep only the specified number of MiB of data on disk. Old blocks get discarded as new blocks come in.

You can give any value you want, but anything below about 3000 MiB provides little value to the network.

Be cautious choosing this number. If you're starting a node for the first time, you'll still have to download the entire blockchain, all of history, before your Core can discard everything too old to stay in the limit. Similarly, if you *increase* this limit, your node will have to redownload everything, just to get the new blocks and set the new limit.

Should you use this mechanism? It depends on how low you want to set things. If you're just running a node for yourself to keep your own wallet up to date, a small limit is fine. You won't provide much value to the network, but if your node is online only a few minutes every day or week, you wouldn't provide much value anyhow.

If you're planning to run your node most or all of the time, it's better to err on the side of having more storage space and pruning as little as you can get away with. 3000 MiB isn't a lot of blocks. 30,000 MiB is a lot better.

Know that a lot of the network burden comes from serving *older* blocks. This doesn't necessarily mean blocks from 2013 and 2014; if many of the nodes on the network only keep a week's worth of data, and if you only bring up your node once every two weeks, you'll put additional load on fuller nodes. The *frequency* with which you sync your node to the current time is important in these calculations.

Limit Network Usage

Unless you have a fiber-optic cable as thick as a professional swordfighter's forearm running to your computer, you probably have bandwidth limits. Maybe you pay an extra penny per packet, or maybe you'll start getting weird popups from your ISP for every terabyte you download[13]

The `maxuploadtarget` configuration option[14] allows you to set a maximum threshold of MiB your node can send to the network in every 24 hour period. The default value of 0 enforces no limit. If you can send 100 GB of traffic every month, then you want to send no more than about 3000 MiB per day (100 GB *

---

[12]Hold on; this will make more sense in a moment.

[13]Your author saw this after he mirrored the entire JavaScript npm repository twice in two days.

[14]Currently only available from *dogecoin.conf* or the command line, but watch the release notes for newer versions!

1000 = 100,000 MiB give or take, divided by 31 days in the longest months equals 3225.8, give or take).

When you hit your daily limit, your node won't send any more data. It will still receive data.

Given the choice between limiting disk space or limiting bandwidth, you might be better off limiting outgoing traffic. Once you buy hard drive space, its yours for the life of the drive, while bandwidth overage charges return every month. Furthermore, the more nodes that can serve old blocks, the easier it is to bring new nodes (even pruned nodes) up on the network.

Limit Network Connections

While the *fastest* possible network would allow every node to connect to every other node, that's infeasible. It's more important for every node to be at most a couple of nodes away from every other node; this means any information can flow through the entire network quickly. Even if you have a good upstream connection, you may want to limit your node's exposure to the network.

Depending on your situation, you may want to ease into the network. For example, if you have a satellite Internet connection, or your node is in a weatherproof box in the middle of a pasture, or you're currently stationed in the Galapagos Islands for a six-month research project, you may want to preserve bandwidth for other people throughout the day. It's smart to be kind, after all.

The `maxconnections` configuration option and its companion RPC command `setmaxconnections`[15] allow you to limit your Core node to no more than the specified number of connections to peers. All nodes have a minimum threshold, so you will not depend on any single node from which to *receive* data. This command limits the number of *additional* connections your node might otherwise make.

Why is this important? Because the network as a whole is trustworthy, even if you cannot trust any given individual node. The more nodes from which you get data, the easier you can trust the accuracy of that data. The more nodes to which you send data, the easier they can trust the accuracy of that data.

Setting this limit is potentially less risky to the network than setting the other two limits; you can still send data anytime throughout your connection, but you're reducing the number of simultaneous connections from your node to the network.

---

[15]New in 1.14.6!

Understand the Risks

With any of these limits, you must balance two risks: your exposure, especially to cost, and your benefit to the network as a whole.

The best thing you can possibly do is run multiple full nodes, geographically and network-ically distributed throughout the world, running the latest software and always available to validate, record, and broadcast transactions and blocks. Not everyone can do that.

The second best thing you can do is to run a node that's as full as possible and is available as much as possible. These tuning options give you the ability to configure your availability to meet your needs and capacity.

  # Tip #17   Ban Bad Actors

Even in the world of funny, dog-themed currency, not everyone is always 100% trustworthy. Some people run scams (Identify a Scam, pp. 23). Others try to cause others grief. Sometimes a bug or misconfiguration even makes otherwise trustworthy people temporarily untrustworthy.

You deserve safety and peace of mind. If you're running your own node, that means you deserve the ability to control the other node connections you allow into your node. In practice, this means two things: either explicitly allowing certain connections or forbidding others.

How Dogecoin Core Manages Network Connections

Your Core node connects to other nodes to send and receive network, block, and transaction information. External nodes fall into one of four categories:

- Whitelisted, where you explicitly allow them to connect to your node

- Normal, where the network helps your node automatically discover and connect to other nodes

- Unconnected/undiscovered, where these nodes exist but your node hasn't connected to them yet (or at all)

- Banned, where your node will reject information from these nodes temporarily or permanently

Most of the time, most nodes on the network are unconnected, undiscovered, or connected as normal. The longer your node has run, the more likely it's connected at least once to any other node in the network. Of course, this depends on your settings and their settings; if you're in Iceland, running a node from geothermal power underground, you're more likely to connect to peers in northern Europe than you are nodes in southern Argentina or Antarctica, if only because the network distance between these nodes is larger.

That leaves two other node categories, each of which require some degree of manual intervention. You have that power.

Allow Specific Connections

If you know certain other nodes are completely trustworthy, perhaps because they're run by close friends, family members, or even you yourself, you can add them to a whitelist in your *dogecoin.conf* file. You can specify either an IP address (IPv4 or IPv6) or a network (again IPv4 or IPv6). For example, if you want to whitelist your brother's node on IP address 10.0.0.101 and your daughter's network at 192.168.2.0/8, you might add these lines to your configuration file:

```
whitelist=10.0.0.101
whitelist=192.168.2.0/8
```

You can also use the -whitelist command-line argument when you launch your node, but configuration file changes are more durable.

This configuration option means that your node will always relay any transactions from these nodes and will never ban these nodes automatically.

Forbid Specific Connections

Dogecoin Core itself monitors node behavior. When it sees something suspicious from another node, it may automatically block connections from that node for a period of time. Suspicious behavior includes sending malformed blocks, sending or requesting too much information, repeatedly connecting or disconnecting, and more[16].

You can use the same mechanism to ban a node or an entire network with the setban RPC command. This command requires two mandatory arguments and supports two optional arguments. The first argument is a single IPv4/IPv6 address or netblock. The second argument is either add to add a ban or remove to remove a ban for the address/network. An optional third parameter lets you specify the number of seconds until the ban expires, unless you provide the optional fourth parameter of true, in which case the third argument is the number of epoch seconds at which the ban expires.

---

[16]For more details, look at the file *src/net_processing.cpp* in the Dogecoin Core

The default ban duration is 86,400 seconds (one day, give or take any daylight saving time shenanigans).

You can change this default value with the `bantime` configuration option, either set in *dogecoin.conf* or specified on the command line.

Durable Peer and Ban Tracking

The Core maintains peer information in two files: *peers.dat* and *banlist.dat*. The former contains information about known peers and the latter contains information about banned node addresses and networks, along with their expiration information. These files are normal and helpful, but the Core can run without them[17].

Whenever you make an explicit change to the banlist with `setban` (either adding or removing a ban), the Core will *immediately* write a new version of *banlist.dat*.

Pre-emptive Blocking

While this specific issue hasn't happened yet, it's possible to imagine that the health of the Dogecoin network as a whole could be at risk from otherwise-trustworthy nodes that have a bug rendering them temporarily hazardous. For example, imagine someone discovers a bug in Dogecoin version 1.11.111.1111[18] where nodes at this version occasionally send messages 10,000 times instead of one. This could have the effect of denying service to peer nodes.

Although the Core *should* detect this situation and consider these nodes misbehaving and ban them temporarily on your behalf, imagine for the sake of example that this somehow doesn't happen.

While you're waiting for the node operators to update to version 1.11.111.1112 or later[19], you *could* write a small script to make the `getpeerinfo` RPC call every minute or so, look for any nodes with a `subver` of the vulnerable version, and call `setban` for their addresses to refuse connections for hours or days.

Again, this is a hypothetical situation to consider only in extenuating circumstances, but it demonstrates a possibility that may happen for which you can preemptively prepare.

Of course, you should also be cautious that a truly malicious node could lie about its version in order to evade such a solution, but the risk of that is always present, and existing monitoring and blocking mechanisms should always work.

---

[17]This doesn't mean you should delete them; just that the Core regenerates them as needed.

[18]This version does not currently and will probably never exist. Any cryptographic news service that claims this bug exists hasn't read this footnote.

[19]Again, this is not a real release.

Understand the Risks

If a misbehaving node is *really* misbehaving, perhaps by trying to scan other network services on your machine, continuing to send requests, using up your resources despite your strong hints that you're just not interested, you may have to use more powerful methods, such as blocking all connections from that node's IP address or network at your firewall, via security software[20], or through your Internet service provider.

This is a powerful mechanism to use if you need it, but it has the drawback of needing management and oversight. Truly bad actors have the habit of attacking from one machine, network, or site then hopping to others to evade bans. If you find yourself targeted by persistent and durable attacks, consider blocking all connections except for a handful of trusted, whitelisted nodes.

This raises one other question: in a distributed, trustless network powered by *consensus*, should you block other nodes at all? The answer depends in part on your philosophy and the security level you personally consider important (see Pledge to Do Only Good, pp. 230). While the design of cryptographic algorithms, Merkle trees, transaction validation, and other mechanisms help everyone verify every bit of information possible, it's up to the network as a whole–meaning all of the nodes, miners, users, and other participants of the network–to agree collectively on what's true, what's false, and what's harmful.

A node that continually broadcasts incorrect data can harm the network just as a node that continually broadcasts malformed or unhelpful data. By not relaying untrusted or unverified data, you protect other parts of the network, just as other nodes should be protecting you. This doesn't mean you should aggressively block any node that doesn't look 100% okay. The features and behavior of the network can and will change over time, after all. Yet the advice to trust but verify applies just as well here. Nodes you can't verify are nodes you can't trust.

---

[20]On Unix-like systems, the utility `fail2ban` works well.

## CHAPTER 3

Network Services with Dogecoin

Dogecoin is more than just a network of nodes, a funny dog-themed currency, or a series of long cryptographic numbers. It's a low-friction payment and tipping system that can integrate with the wider Internet. This chapter explores several ways to do so, both as a publisher and a consumer.

  ## Tip #18   Add a Wallet Address to a DNS Record

If you have a website, like https://ifdogethenwow.com/, how can people find you and send you a tip or a payment for something awesome you've done? You need some way to associate a Dogecoin address with an Internet property.

Fortunately, the Internet already has a well-understood mechanism to add this kind of metadata to a domain name. It's the same way any device can turn that domain name into an IP address or addresses to look up a web page, send email, make an SSH connection, or do other things: DNS, the domain name system.

At the inaugural Dogecoin hackathon[1], Timothy Stebbing showed off this idea[2].

How to Add a TXT Record

Assume you have a website. How do you allow people to send you tips? First, generate a new, unique Dogecoin address (see Replenish Your Address Pool, pp. 160, for example). Use the Dogecoin Core, `libdogecoin`, or a non-custodial wallet you trust. Keep the private key safe, as always.

---

[1] https://foundation.dogecoin.com/announcements/2022-09-08-dogeathon-downunder/

[2] . . . and, inadvertently, kicked off the writing of this book. Thanks, Timothy!

Second, take a domain name you control. Maybe that's `mycoolsite.example` or something else. You've registered this through a domain registrar, and you have the ability to configure DNS.

> **Keep Your Domain Running!**
>
> If you didn't set up your own DNS, talk to the person or people who set it up for you. If you edit this on your own and make a mistake, you could break your website, email, or other network services associated with that domain.

Go into your DNS configuration and add a 'TXT' record. This is an arbitrary text string associated with your domain name and propagated through the globally-accessible, cached DNS system. For now, stick with the defaults of *which* DNS entry, TTL settings, et cetera.

The contents of this new record should be the literal string 'dogecoin:' (yes, include the colon) plus the new wallet address you generated, with no spaces in between. This conforms to BIP-21[3], so other applications can use this. Hold onto that thought.

Save the configuration and publish it. In a few minutes, when your DNS changes propagate to the rest of the Internet, you'll be ready to go. Anyone will be able to look up your domain name, find this new record, and do something with this wallet address.

> **Specific DNS Configuration Details See...Elsewhere**
>
> Looking for a visual explanation of how to do this? The process varies depending on your domain registrar and DNS host. Again, this has the potential to render your domain *temporarily* unusable, so be cautious, read the documentation, and ask for help if you need it.

Understand the Risks

At this point in the book, you should be a little bit wary of the words *anyone* and *something*, because publishing an address like this means anyone can do *anything*

---

[3] See https://github.com/bitcoin/bips/blob/master/bip-0021.mediawiki

with both the address and the knowledge that the address is associated with the domain name.

DNS entries often contain contact information for the person managing the domain. While many registrars allow domain security and privacy to hide your personal information, some don't. Before you add this record, check your DNS settings with a DNS lookup tool like the `whois` command-line tool or a reputable website. If you see your name, address, phone number, and email address available, then so do other people.

You *can* of course add this record even if you have contact information recorded, but be mindful of the fact that any transaction to and from this address can be associated with the domain contact permanently.

You should also be aware that DNS as originally designed has potential security flaws, including spoofing. If you're concerned about this potential flaw, make sure you've configured DNSSEC correctly. Otherwise, someone could pretend to be you, change this DNS information, and swap their Dogecoin address for yours.

Finally, if you fail to renew your domain or someone snatches it up out from under you, they can swap their address for yours in their configuration, in the same way that they could swap their website for yours if you lose control of your domain name.

What Can You Do With This?

Apart from being an interesting forehead-slapper of a trick[4], the value of this trick is in the integrations it enables.

Imagine a web browser extension that could show you a Dogecoin tip address whenever you visited, if the site owner had one configured. If you like the site, throw a few Doge their way, right from your browser or mobile device.

Or imagine we repurposed this record to point to a text file or structured data file on the server, served over HTTPS, that gave access information, purchase data, or other suggestions, like "If you send 10 Doge to this address, you'll get 24 hours of access" for a news or other subscription site.

Consider also an alternate approach. `ifdogethenwow.com` has, as of this writing, the main domain and a well-known[5] subdomain of `blog.ifdogethenwow.com`. Any and every subdomain could have its own TXT record with a unique wallet address. An enterprising shibe could start a site called `shibetips.whatever`

---

[4] In the sense that, "Oh yeah, you could totally do this!"

[5] Okay okay, hear me out though!

and, for a small fee or out of the goodness of their hearts, allow users *optionally* to associate an address with a subdomain. As always, keep in mind the security and privacy implications, but think of the possibilities.

  ## Tip #19   Write a JSON Block Explorer

If a blockchain is just an append-only database shared by multiple machines across a network, then any meaning anyone put onto the data of that chain is a series of models humans apply to raw data. While that could be said of anything computers do, it's especially true of blockchains, because a blockchain is just a bunch of numbers.

If that's too abstract for you, consider this: we talk about blocks, transactions, and addresses not because the blockchain *requires* that we talk about these things but because it helps us humans figure out what's going on and build things for other humans to use.

A piece of software called a *block explorer* can turn that semi-structured data from a blockchain into data that machines can more easily process (even if they don't understand the blockchain protocols)–and also data that humans can understand. After all, even though a transaction is just a blob of hexadecimal code, it's easier to figure out when you can see where the inputs came from and where the outputs go. A block explorer translates block, transaction, and address data into a form that's meaningful to humans and other machines.

Multiple services provide this data via API calls. Some let you use their data for free. Others charge money. The drawback in all cases is that you have to trust someone else for this data. Could you tell if they're subtly wrong? What if they cut you off? What if the service goes away? What if they sell your data?

Those risks might be worth taking, but you don't have to take them. You can control your own data.

Write Your Own Block Explorer

Let's focus on transactions and addresses to demonstrate this possibility. Dogecoin Core's RPC mechanism (see Enhance RPC Calls, pp. 80) provides a way to get this data, and creating, using, or re-using a wrapper around core RPC calls can make this task easier.

At one point, the `blockchain.info` block explorer provided an API which served JSON transaction data from `https://blockchain.info/rawtx/...` and address data at `https://blockchain.info/rawaddr/...`. The former included things like the number of confirmations, block height, time, and inputs and outputs. The latter included unspent amounts and all transactions with outputs to the address.

Let's start with transactions.

Serving Transaction JSON

The API URL pattern makes it easy to know what this API has to do: provide a transaction hash and then produce the results. To continue the idea of RPC wrapper, let's reuse the `Finance::Dogecoin::Utils::NodeRPC` code described elsewhere and combine it with the excellent Perl web framework Mojolicious[6] in its lightweight version. You can use whatever code you want, of course.

The interesting part of the API looks like this:

```
get '/rawtx/:txhash' => sub ($c) {
    my $txhash = $c->param('txhash');
    my $txdata = $c->node_rpc->
        call_method( getrawtransaction => $txhash, 1 )->{result};

    $c->render( json => $txdata );
};
```

If you've never read Mojolicious code before, that's okay. The most important line is the third line, which calls a method called `call_method` on a Node RPC object. This method in turn makes a `getrawtransaction` RPC call against a Core node and returns the result. `getrawtransaction` takes a transaction hash and an optional parameter to include full data and returns a JSON blob of transaction information.

There's a flaw or weakness in this code, however. If you make this RPC call against an arbitrary transaction that doesn't affect an address in your wallet, you'll get an error response saying that this transaction hash isn't in your mempool. In that case, you need to do one of two things. First, you could restart your node with the `txindex` configuration option (Index All Transactions, pp. 192). Second, we could rewrite this code to expand the transaction ourselves:

```
get '/rawtx/:txhash' => sub ($c) {
    my $rpc    = $c->node_rpc;
    my $txhash = $c->param('txhash');
    my $raw_tx = $rpc->call_method(
        gettransaction => $txhash
```

---

[6]See https://mojolicious.org/ for more details.

```
    )->{result};
    my $txdata = $rpc->call_method(
        decoderawtransaction => $raw_tx->{hex}
     )->{result};

    $c->render( json => $txdata );
};
```

The second and third lines have changed; now the code calls `gettransaction`, extracts the `hex` field from the result, and passes that to `decoderawtransaction`. While this probably should become its own all-in-one RPC command, or at least something proxied to it by the Perl library used here, it serves its purpose in the API and demonstrates a way to go from a transaction's hash to its full data.

Serving Address JSON

Address information is also available through RPC commands, but they don't make it easy. The `listreceivedbyaddress` command returns a list of all known wallet addresses and their current unspent balances as well as all transactions for which the address was an output. Unlike transaction data, it's not as easy to get at *any* arbitrary address, but there's still useful information here.

```
get '/rawaddr/:address' => sub ($c) {
    my $address = $c->param('address');
    my $rpc     = $c->node_rpc;
    my $received = $rpc->call_method('listreceivedbyaddress')
                        ->{result}

    my $addy_data;

    for my $addy (@$received) {
        next unless $addy->{address} eq $address;
        $addy_data = $addy;
        last;
    }

    return $c->render( json => [] ) unless $addy_data;

    $addy_data->{txs} = [];

    for my $txhash ($addy_data->{txids}->@*) {
        my $raw_tx = $rpc->call_method(
            gettransaction => $txhash
        )->{result};
        my $tx     = $rpc->call_method(
            decoderawtransaction => $raw_tx->{hex}
        );
        push $addy_data->{txs}->@*, $tx->{result};
```

56

```
    }

    delete $addy_data->{txids};

    $c->render( json => $addy_data );
};
```

This batch of code does two things. First, it finds the information for the currently requested address in the data returned from `listreceivedbyaddress`. Second, it loops through each transaction ID for that address and get the full transaction information, just as the previous transaction-handling route did.

While this isn't a full general-purpose block explorer because of the address limitation, you can solve this with some clever thinking (hint: see Watch Wallet Addresses, pp. 157).

The Rest of the Code

The remainder of the code looks like:

```
use v5.036;

use Path::Tiny;
use File::HomeDir;
use Mojolicious::Lite -signatures;
use Finance::Dogecoin::Utils::NodeRPC;

helper conf_dir => sub ($self) {
    state $conf_dir =
        path(File::HomeDir->my_data)->child('dogeutils')->mkdir;
    return $conf_dir;
};

helper auth_file => sub ($self) {
    state $auth_file = $self->conf_dir->child('auth.json');
    return $auth_file;
};

helper node_rpc => sub ($self) {
    state $node_rpc = Finance::Dogecoin::Utils::NodeRPC->new(
        user      => $ENV{DOGEUTILS_USER},
        auth_file => $self->auth_file,
    );

    return $node_rpc;
};

get '/rawtx/:txhash' => sub ($c) { ... };
```

```
get '/rawaddr/:address' => sub ($c) { ... };

app->start;
```

Most of this is helper code to allow you to use the authorization file set up for the RPC proxy (again, see Enhance RPC Calls, pp. 80) and to create the RPC object. Pay close attention to the use of the DOGEUTILS_USER environment variable (see Authenticate RPC Securely, pp. 73) to provide the name of a user for the RPC commands.

Run this code with:

```
$ morbo bin/dogeblockserver
```

...and you should be able to navigate to http://localhost:3000/ and start making these raw transaction/address calls.

Understand the Risks

What's risky here? If you don't go further and figure out how to index all addresses *or* add some kind of HTTP authorization, anyone who can make JSON HTTP requests against this server can figure out which addresses belong to the wallet associated with the Core node.

Furthermore, for this to work at all, you have to have a wallet associated with the Core node, so make sure you protect that wallet to every extent possible. It may make sense to use only watch-only addresses so you don't have any private keys accessible from that wallet. It'd still be unpleasant for someone to get your data, but they can't spend without your private keys, so your damage is slightly less.

> **I'm Asking Again, Why Perl?**
>
> You can do this in any language you like, and it's important to have multiple implementations in multiple languages, so your author chose something that was fast for me to write, entirely under my control, and not so popular or complex that people would be intimidated to reimplement it in their own favorite languages.

What Can You Do With This?

Now that you have a way to get address and transaction data out of a Core node into a machine-readable format *and* you can do so over any network that supports

HTTP, you can build all sorts of applications and tools. You can go further and add block navigation and exploration support. You could put a nice HTML/JavaScript interface on this data and make calls back into the system to click through blocks, transactions, and addresses.

You could visualize the entire blockchain in 3D or other graphical formats.

Of course, you can also make the server more robust. Perhaps it should support HTTPS, so putting the application behind a reverse proxy to terminate TLS connections can make it more secure. Maybe it could support some sort of token or authentication system to limit the types of queries specific users could ask.

Because transaction and block information is fixed and immutable, you could even cache query results to avoid unnecessary round trips between the server and the Core node. Caching address information is trickier, because every new block processed could consume an address's input or add an output to an address, but you could set cache headers of 60 seconds and provide good responsiveness.

None of this code is so complex you couldn't port it to another language; deploying an all-in-one binary with Golang or Rust could make a trivial but useful addition to a node you run in the cloud or on a network you control somewhere. The important thing is that this data is entirely under your control, with no access limits, API permissions, or waiting on someone else to add or change features.

  # Tip #20   Explore the Memepool

You know about addresses, transactions, and blocks, but there are more things under heaven and earth than that. You may wonder "what happens to a transaction that isn't yet mined into a block". The answer is slightly complicated! When you create a transaction and submit it to the network, it gets broadcast through the network until it reaches a miner that mines it into a block.

Transmission through the network and mining all depends on the transaction being valid: the transaction's inputs are valid and haven't been spent, the transaction's signature is valid, and the transaction's fees are appropriate. If enough nodes and the miner validate all of this information, the transaction is available to be mined into a block. Until that mining happens, the transaction is pending.

The entire set of pending transactions is the *mempool*. You can (and sometimes should) inspect the mempool to understand what's going on, especially if you have a pending transaction. You can also inspect it for amusement.

Mempool or Memepool?

Given that Dogecoin is a friendly, dog-themed cryptocurrency designed for fun, wouldn't it be more fun to call the mempool a memepool? It's probably too late for that (and Very Serious Cryptocurrency People will raise their eyebrows and shake their heads[7]), but we can still have a good time.

What if there were a way to tie the mempool to memes?

The `getmempoolinfo` RPC command produces basic information about the current state of the mempool:

```
{
  "size": 823,
  "bytes": 476227,
  "usage": 1458688,
  "maxmempool": 300,000,000,
  "mempoolminfee": 0.00000000
}
```

Given this information, we know that there are currently 823 transactions in the mempool, adding up to 476 thousand bytes in size. This costs the Core about 1.459 thousand bytes of memory, with a maximum memory size allowed of 300 million bytes. Finally, the Core estimates no minimum fee to process the transaction[8].

At the moment, this information is interesting but not essential for anything other than the possibility for a silly pun. Let's pick the `size` parameter; when the size changes, let's do something interesting.

Turn Mem to Meme

Assume you've used your RPC access mechanism (see Enhance RPC Calls, pp. 80) to fetch the current mempool size. This gives you a semi-random number you can use in other ways.

How do you turn a number into a meme? One of the easiest options at the time of this writing is `imgflip.com`[9]. If you fetch the url https://api.imgflip.com/get_memes, you get back a JSON data structure with a list of memes (actually meme templates, but hold that thought).

Given the Dogecoin mempool size and a list of memes, you can use the size as an index into the list to pick a meme. In Perl, the code would look something like:

---

[7]They always do that anyway, so don't let that stop us from having a good time.

[8]Don't rely on this number; use the fee recommendations for Dogecoin 1.14.6 instead.

[9]See the API at https://imgflip.com/api.

```
use Modern::Perl '2023';
use Mojo::UserAgent;

exit main( @ARGV );

sub main {
    my $tx_count = fetch_mempool_size();

    my $ua    = Mojo::UserAgent->new->max_redirects(3);
    my $json  = $ua->get(
        'https://api.imgflip.com/get_memes'
    )->res->json;

    my $memes = $json->{data}{memes};

    my $meme_count = @$memes;
    my $idx        = int($tx_count % $meme_count);

    say $memes->[$idx]{url};

    return 0;
}
```

This code is straightforward[10]. The first lines load two libraries, `Modern::Perl` to set language defaults and `Mojo::UserAgent` to make web requests.

The `main` function first calls a function to fetch the current mempool size, the implementation of which you can find alluded to in other tips.

The next three lines construct a user agent ($ua) object, make an HTTP GET requests against it, then access the `data` and `memes` members of the returned object. See the imgflip API for more details about this data structure.

`@$memes` gets the number of items in the array of memes returned from the API call.

That leaves one interesting line of code, using Perl's modulus operator (see Roll Over Your Odometer, pp. 5) to reduce the value of `$tx_count` to a number between 0 and the current number of memes returned by the imgflip API. If there are 12 memes in the response and 25 transactions in the mempool, the resulting value will be 25 % 12 or 1.

Finally, the `say` line fetches the meme from the array at that index, grabs the `url` of that meme, and prints the URL to standard output. You can do more with this; customize the output to your own preferences!

---

[10]Especially if you understand Perl.

What Can You Do With This?

There aren't many risks from using this code, as long as you're accessing RPC securely (see Authenticate RPC Securely, pp. 73). Be cautious accessing any API remotely, however; DNS records could change and someone could send back a malicious payload. If you make any outgoing network connection from a machine hosting a node, lock down your node very carefully and monitor your system to make sure nothing untoward happens.

The example code here performs very little error checking. In a production setting, be very cautious.

  Tip #21   Remember Updated Nodes

Any node that participates in the Dogecoin network has to adhere to a few rules about following network consensus, and transmitting good data: basically *behaving well*. Good nodes also help keep the network robust, by transmitting necessary data (such as block history and pending transactions) and enforcing network guidelines (including block size, difficulty, and fees).

There's no requirement your node has to be a Dogecoin Core node, but if you're running a node, you're better off running the latest stable software–and you're better off if more and more nodes on the network are doing the same, or at least running software with similar rules.

How do you know what they're running and which nodes are good, whatever good means?

Peer Into Your Peers

A Core node tracks a lot of data about the peer nodes it's connected to. This is available via the `getpeerinfo` RPC command, which returns data including:

```
[{
  "id": 64,
  "addr": "65.xxx.xx.xxx:22556",
  "addrlocal": "73.xxx.xxx.xx:34946",
  "version": 70015,
  "subver": "/Shibetoshi:1.14.6/",
  "startingheight": 4817965,
  ...
}, ...]
```

There's a lot more information available, but even with only this much shown, you can do interesting things. The `id` is a unique identifier for the peer. `addr` shows the peer's IP address, whether IPv4 or IPv6. addrlocal shows how the peer connects to your node (IP address and port). version shows the version of the protocol in use, and you want to see 70015 for lots of reasons[11]. subver shows the Core version in use as well as any user comment the peer added (see Set Your Node Comment, pp. 38). Finally, startingheight shows the oldest block height the peer has available.

What makes a node good? That's up to you to figure out. One place to start is "running the latest software" and "has a lot of blocks available".

Filter Good Peers

Start with the easy question. Which peers you're currently connected to are running the latest Core version? At the time of this writing, 1.14.6 has been out for a while, so most peers should have updated by now. The getpeerinfo command's output is a big blob of JSON with a list of objects, so use jq to extract data from the JSON. Assuming you've stored the output of getpeerinfo into a file named *peers*, you might write:

```
$ jq '.[] | select(.subver | contains("1.14.6")) | .addr' < peers
```

This will give you a list of all of the IP addresses of all of the connected peers using the latest stable version of the Core. If you prefer to find nodes with more blocks, filter on startingheight instead:

```
$ jq '.[] | select(.startingheight < 123456) | .addr ' < peers
```

Of course, there may be a lot fewer peers with older blocks than run the newer version; storing the entire blockchain takes up some space after all (see Run a Node, pp. 34).

What Can You Do With This?

There aren't many risks from using this code. Depending on how many nodes you connect to (see Know Your Limits, pp. 44), you probably only see a subset of the network. If you use this tip to populate a series of nodes you want to try to connect to (see Ban Bad Actors, pp. 47), remember that *you* are as much a part of the network as anyone else who runs a node. While it may be safer to connect to updated nodes, it's better for the network for you to connect to nodes without all of the blocks, especially if you're running a full node yourself.

---

[11] Reading the source code will help, so if that's not something you want to do right now, remember to ask someone you trust about it in the future.

There's no reason you have to use Core version, blocks available, or any other stats as the basis for what you consider "good" or memorable. Any other data you have available on peers is fair game.

Also remember that peers come and go. You may want to run this tip once an hour and collect good peers over the course of a day or a week. Good nodes that are often online are worth connecting to again–but that's only an *enhancement* to your own node. Let the protocol and the network make connections and add enhancements and suggestions only when it makes your life (and the network) measurably better.

Finally, remember that you're not the only one who can see this data. Your node broadcasts this information to the network, and your information is available. Respect the privacy of your fellow node operators, and think very carefully if you publish any potentially identifying information in a public place. Even an IP address could give up the privacy of a fellow shibe. That's why they're redacted in this tip's examples.

  # Tip #22   Post to Discord

If you turn your head and look at Dogecoin blocks and transactions from a slightly different angle, you can think of them as a bunch of data published to the Internet on a regular basis. In one sense, they're payment events with a bunch of data attached to them: $X Doge moves from address $A to address $B, in a mathematically verifiable way.

Any time you see an event published to the Internet with a bunch of data attached, you can ask yourself "What can I do with this?" In the case of Dogecoin blocks and transactions, the answer is "a lot". While other tips in the book talk about controlling physical devices (all of Manage a Dogecoin Arcade, pp. 301 and Control Your Jukebox, pp. 221), you can do other things as well, such as sending yourself a notification on your phone or publishing a message to a chat room.

Inside a Webhook

Let's start with Discord, a chat service which allows you to create servers and channels for friends, communities, and more. It has a couple of good Dogecoin and Dogecoin-related servers, such as the Dogecoin Discord server[12].

---

[12] See https://discord.gg/dogecoin.

Many online services provide a feature called a *webhook*, which is a way to send specifically-formatted data to a URL to perform some action on your behalf. Discord is one such service. If you have access to configure a server or a channel within a server, you can sign up for a webhook which allows a program to send a message to a specific channel.

In other words, a webhook is an entry point into a service which uses pre-authorized credentials to perform an action on your behalf.

Inside a Discord Webhook

You've probably already guessed that Discord supports webhooks without much configuration *and* it has a good introduction to webhooks[13] to help you create and configure your own.

To use this tip, you need your own Discord account and access to a server or channel. You can follow along with this tip with a different type of service for which you have similar access, though the details will be different.

After you create your webhook, you'll end up with a URL which looks something like `https://discord.com/api/webhooks/<number>/<string>`. Keep that safe and secret; anyone who has that URL can send messages to your Discord server/channel. You probably don't want that to happen.

Activating Your Webhook

Every webhook is a little bit different. A Discord or Slack or other messaging webhook will have a different format than a webhook for a deployment or testing service. While Discord lets you format messages with plenty of complexity and formatting, the simplest way to send a message is to use an HTTP POST request with a JSON body, with two keys, `content` and `username`:

```
$ curl -X POST -H "Content-Type: application/json" \
  -d '{"content": "Much wow!", "username": "dogecoin"}' \
  https://discord.com/api/webhooks/<number>/<string>
```

Fill in your URL appropriately, and you should see a message appear in the appropriate channel in your server.

Putting it All Together

Now that you've configured and tested a webhook, you can do something more interesting: maybe send a message each time you receive a transaction, share details

---

[13] See https://support.discord.com/hc/en-us/articles/228383668-Intro-to-Webhooks

about every new block that comes in, or something else. For example, here's code to send a message to your secret channel every time your Core receives a wallet transaction (see Act on Wallet Transactions, pp. 194):

```bash
#!/bin/bash

txid="$1"
tx_json=$(dogecoin-cli gettransaction "$txid")
tx_amount=$(jq -r .amount <<< "$tx_json")
content="Much wow, received a transaction of $tx_amount Doge!"

curl -X POST -H "Content-Type: application/json" \
  -d '{"content": "$content", "username": "Doge TX NotifyBot"}' \
  https://discord.com/api/webhooks/<number>/<string>
```

You can make this a *lot* more robust, per the other tips, but this is a good start at tying everything together.

Understand the Risks

As mentioned earlier, your webhook URL is a secret. If you write it down on a piece of paper and accidentally leave that paper in a public place, someone could use that to post on your behalf, to nefarious purposes. This gets worse if you or someone else has *another* webhook or automated process which acts on the first webhook.

You can use two common mitigation strategies to reduce your risk. First, use a secure configuration system such as a vault or secret manager to store the webhook URL. Second, consider rotating the webhook on a regular basis. Even the exercise of thinking through how to rotate the webhook will help you to understand the implications, in case you have to do it in an emergency.

  Tip #23   Serve Doge Data from DNS

A previous tip about adding wallet addresses to a DNS record (see Add a Wallet Address to a DNS Record, pp. 51) opened the door to serving static or at least semi-static data from DNS. There's plenty more where that came from!

For example, retrieving data from a node under your control requires configuration and security (see Getting Data from a Local Node, pp. 73). While this remains true, using an existing network service such as DNS to serve this data can simplify your life, especially if the alternative means exposing RPC or other services to the Internet—or pushing you to rely on someone else's potentially untrustworthy copy of the data.

66

Generating Data from DNS Queries

Kailash Nadh's DNS Toys[14] is a great example of a useful principle behind DNS: a query is a kind of request, and a response is a kind of answer. While the original purpose of a DNS query is to resolve a domain name into an IP address for serving HTTP content, responding to email queries, et cetera, there's no reason a *custom* DNS server cannot respond to non-domain-name queries with custom data.

In other words–in Kailash's examples–a DNS query for `berlin.time` does not represent an actual domain name but instead gets interpreted as a request for the current time in the city of Berlin.

For this to work, the querant has to understand the format of the query (city name dot time, in this case) and the server has to understand both the query format (all dot time requests are time requests) and the data format to return. A general purpose DNS server will respond with "There's no such domain name", so you need a custom DNS server.

Responding to Custom Queries

Responding to custom queries is easy with the right DNS server. For example, your author knew about the Perl library `Net::DNS::Nameserver`, so he was able to write a responder in a few minutes. You can use whatever language or toolkit you like; this is only for example purposes.

Start with something easy, such as a query for the oldest and newest block heights that a Dogecoin Core node can serve. Why is this useful? Perhaps you have a pruned node (see Know Your Limits, pp. 44>) and you want to know how far back you can go to retrieve data. Perhaps you have a regular process which needs to examine newer blocks and transactions and you want to know if you're up to date.

In this case, a good format might be `param.block.doge`, where `param` is either `first` or `last`. This pattern is easy to parse and should be easy to extend.

The responder should be able to respond to queries of this form with a TXT record. For the sake of argument, consider returning a semi-structured string something like "First block height is 1" or "Last block height is 5038774".

`Net::DNS::Resolver` allows you to specify a custom function to act as a reply helper, so the body of the Perl code will look something like:

```
sub reply_handler( $qname, $qclass, @rest ) {
    state %actions = (
```

---

[14]See https://www.dns.toys/.

```
        block       => \&handle_block_query,
        transaction => \&handle_tx_query,
    );

    my @results;

    my ($param, $type, $doge) = split /\./, $qname;

    if ($doge eq 'doge' && $actions{$type} {
        @results = $actions{$type}->( $param, $qclass, @rest );
    }

    @results = ( 'NXDOMAIN', [] ) unless @results;

    return @results, [], [];
}


sub handle_block_query( $param, $qclass, @rest ) {
    my $height = $param eq 'last' ? 1 : 5038774;
    my $txt    = ucfirst $param . " block height is $height";

    return 'NOERROR', [ make_txt_record( block => $txt ) ];
}

sub make_txt_record( $name, $txtdata, $ttl = 60 ) {
    return Net::DNS::RR->new(
        name    => $name,
        type    => 'TXT',
        ttl     => $ttl,
        txtdata => $txtdata,
    );
}
```

The first function takes several elements from the DNS query; the most important is $qname which, in this case, contains something like first.block.doge or something entirely unlike that. The function splits the response into three parts corresponding to the three parts in the query. If the last part is not doge, the function returns an error response which means "The domain does not exist".

For the case when the last part *is* doge, the function calls a helper named handle_block_query. Here there are two hard-coded values for the last block and everything else, assumed to be the first block. This function itself calls another helper to make a TXT record object which the Perl library can translate into the appropriate output.

This make_txt_record helper is not strictly necessary, but it allows further customization. For example, it includes a parameter to set the time-to-live (TTL) value for the record. The default is 60 seconds. Any DNS caching between this

server and the client you use to query the server can cache a response for the
duration of the TTL value, reducing network traffic.

To test this service, you need a little more code to run it:

```
use Modern::Perl '2023';
use Net::DNS::Nameserver;

exit run( @ARGV );

sub run( $timeout = 20 ) {
    my $ns = Net::DNS::Nameserver->new(
        LocalAddr    => '127.0.0.1',
        LocalPort    => 5354,
        ReplyHandler => \&reply_handler,
        Verbose      => 1,
    ) || die "Couldn't create Net::DNS::Nameserver object\n";

    local $SIG{INT} = sub { $ns->stop_server };

    $ns->start_server( $timeout );
}
```

Save this file as *dogedns.pl*. Install Modern::Perl and Net::DNS::Nameserver,
and then you can run it from the command line:

```
$ perl dogedns.pl
```

To test it, use the dig command-line tool:

```
$ dig +short @127.0.0.1 -p 5354 last.block.doge
"Last block height is 5038774"
$ dig +short @127.0.0.1 -p 5354 first.block.doge
"First block height is 1"
```

The +short option tells dig to only print the response data, not the remainder of
the query information. You can remove it to see more information. @127.0.0.1
queries the local host (set in the run function) and -p specifies the port number
(also set in the run function).

Serve JSON From Custom Queries

Serving hard-coded block heights isn't interesting for long. It'd be more useful to
serve the actual block heights from the node. This requires a little more work, but
not much (see Enhance RPC Calls, pp. 80, for example).

Returning text or block heights themselves is also interesting, but serving structured data–such as the JSON of a transaction itself–is a lot more interesting. What

if there were a query scheme something like `txid.transaction.doge`? Adding that to the handler could be:

```
use JSON 'encode_json';

...

  state %actions = (
      block       => \&handle_block_query,
      transaction => \&handle_tx_query,
  );

...

sub handle_tx_query( $param, $qclass, @rest ) {
    my $tx = get_transaction( $param );

    return 'NOERROR',
      [ make_txt_record( tx => encode_json( $json ) ) ];
}
```

... where `get_transaction` knows how to call the RPC interface to get the JSON of the node. Note two things here. First, there's no error handling (and there should be). Second, the `make_txt_record` helper creates another TXT record with a name of `tx` and a single entry containing the encoded JSON of the transaction. Behind the scenes, the Perl library will format the encoded text such that it conforms to the DNS protocol specification. That means that *consuming* the output is a little trickier, as you can see if you run another `dig` query to see the output.

To consume this data with `jq`, for example, you need to remove quoted spaces to concatenate together multiple TXT entries:

```
$ dig +short @127.0.0.1 -p 5354 tx.transaction.doge |
  sed -r 's/" "//g' | jq ". | fromjson | ."
```

Swap `tx` for the id of a transaction your node can serve. Here, the `sed` Unix utility removes all quoted spaces from the output, which allows `jq` to parse the resulting string as a JSON string. The `. | fromjson | .` part of the `jq` command parses the JSON string *as* JSON, and then prints it out in a more readable format.

If you're consuming this data from your own code, make sure your library or parser handles the output here appropriately. Similarly, if you implement this in a different language, see what kind of output you get from providing long strings as TXT responses.

70

Understand the Risks

Exposing any data from a node (or a machine with access to a node) represents a point of connection to the node. Beware that any exploit of this service could lead to a compromise of the machine where it runs and, from there, the node itself.

Do you want Perl and a stack of libraries running on your machine to serve DNS queries? Perhaps or perhaps not. It might be interesting to port this code to Golang or another single-binary distributable language–though your ability to hack on it may be more interesting than the language itself.

This code runs on an unprivileged port (5354) so it doesn't need superuser privileges to run. It also runs on the loopback interface so that only the local machine can query it. If you want to change either of those properties, be very cautious. If you're not sure what that means or what they imply, ask some trusted friends and consider reconsidering what you want to accomplish.

What Can You Do With This?

This is currently a toy, but it demonstrates where things could go.

You could combine this with the DNS wallet address service to create a registry of names to addresses, even subdividing things further. For example, each user of your system could have their own person.shibes.doge subdomain which returns information including a tip wallet address, contact information, et cetera[15].

If you find that transaction IDs are too lengthy or confusing to use, you could associate labels or other aliases to them, corresponding to labels for addresses.

Any read-only RPC call could be served over DNS. You could even serve (limited) write calls over DNS, but the thought of doing so securely gives even your author pause.

There's more of DNS to explore here as well. Serving blocks and transaction information over DNS has some big advantages.

The transaction response uses the default 60 second TTL setting, but because transactions are effectively immutable after a few block confirmations, you could set a TTL of days, weeks, or months for those responses. Similarly, 60 seconds might be too long for block heights, as blocks should get mined every minute or so but may be mind more frequently.

If you find yourself serving a fleet of nodes or systems making a large number of queries, you could put a caching DNS server between your clients and the

---

[15]Remember to practice good security and privacy hygiene when doing so.

71

DNS server and rely on the intermediary to handle TTL and response caching appropriately.

The biggest problem/advantage to consider is that individual clients and servers and languages and libraries may all handle DNS in different ways. This tip pushes the intent of DNS perhaps further than it's generally used, so perform lots of testing before you rely on it.

Getting Data from a Local Node

When you run your own node, you can do more than mine blocks and keep the network healthy. You get insight into the current, past, and future state of the network.

This is powerful. You do not have to rely on someone else to tell you what's happening, what might happen, or what has happened. All of your data is in your own hands, with no one modifying it behind your back, charging you for access, or keeping track of what you're doing.

Here's how.

  Tip #24   Authenticate RPC Securely

A running Dogecoin node holds a lot of information. If you ask it nicely, it can give you this information: what's in your wallet, what happened in a given block, the current difficulty, the contents of the mempool, and more.

When you read the phrase "your wallet", think about what your wallet represents: not just your keys, but the transactions you've made. Who paid you? Who did you pay? When did this happen? If you think "Wait, I don't want just anyone to that information", that's good. You probably don't.

Securing your wallet and your node keeps your coins *and* your privacy safe. That means you need to restrict access to your node. Fortunately, the Dogecoin Core has a way to ensure that only the people who are supposed to connect to a node can do so. All you have to do is configure it securely.

Understanding Secure Authentication

To connect to a Core node security–for example, to ask it for information or to do things on your behalf–you need to authenticate, or prove your identity, in two ways:

- The node recognizes the user you claim to be

- The node confirms that you are that user

The straightforward approach is to give the node a list of authorized users and some way of confirming the identity of those users. In other words, if the server knows about `ralph`, `nelson`, and `milhouse` and someone tries to connect as `jimbo`, the node can easily deny that connection.

If Jimbo gets smart and tries to connect as `milhouse`, the node also needs to check that Jimbo has permission to connect as `milhouse`. He can "prove" this by providing `milhouse`'s password. If Jimbo tries `bartrules`, `imissmymom`, and `pantsme`, he'll have trouble. If he manages to guess `thrillho`[1], he's in.

For this approach to work, the Core has to know about usernames and passwords.

Add Users to Your Config File

How does the Core know `ralph`, `nelson`, and `milhouse` are all valid users and to reject invalid passwords? The answer is in your *dogecoin.conf* file and a Python program from the Dogecoin Core repository at *share/rpcuser/rpcuser.py*[2].

If you have the Python programming language, run this program like:

```
$ python3 share/rpcuser/rpcuser.py lisa

String to be appended to bitcoin.conf:
rpcauth=lisa:b0a414f0d217c5bdec8db24ff340223$...
Your password:
...=
```

This program generates a random password for you to remember and a string you can add to your configuration file. Keep those two items separate but close. First, grab the line starting with `rpcauth` and add it to your *dogecoin.conf* file, then save the file. Start or restart your node.

Second, take the password and store it somewhere else, securely. Don't write it on your hand; that's what Milhouse did and that's how Jimbo impersonated him.

---

[1] See https://www.youtube.com/watch?v=nbbKsAZatao.

[2] See https://github.com/dogecoin/dogecoin/blob/master/share/rpcuser/rpcuser.py, for example.

You could, of course, modify the code to provide your own password, but the random value you get here is going to be difficult to guess. Bitcoin has an updated version of this code (at least newer than the version in Dogecoin Core 1.14.7; version 1.14.8 updated this code), so look in their repository for a file named *rpcauth.py* for more options when generating passwords. In particular, the Bitcoin utility uses more entropy (see Embrace Entropy, pp. 21) when generating random passwords–giving you more security.

When your node starts, test your authentication by connecting with your preferred RPC mechanism and issuing a command. The `getdifficulty` RPC command is a good test for connectivity. The `curl` binary is a good client to use, because it has nothing to do with Dogecoin or any other cryptocurrency. If a request like this works, you can be confident that you have configured things correctly:

```
$ curl --data-binary '{
    "jsonrpc":"1.0",
    "id":"curltext",
    "method":"getdifficulty",
    "params":[]}' \
    http://lisa:...@127.0.0.1:22555
{"result":8101199.900972591,
 "error":null,
 "id":"curltext"}
```

This example assumes you're running `curl` on the same machine as your Code node. Change the IP address (here, `127.0.0.1`) if not. If you get the authentication wrong, you'll instead receive error output.

How Does This Work?

Run the Python code again and you'll see different results: both the password you use and the string you need to add to your configuration file. The code picks random values for you, with the password it provides and the salted value.

Why does this work?

Neither the configuration file nor the Core store or know your password. The information you add with the `rpcauth` line is two things: a random salt and the *hashed* value of the combination of the salt and your password (see Make a Hash of Fingerprints, pp. 1).

To prove that you're `lisa` or `milhouse` or `nelson` or anyone else the node knows about, you have to provide your username and your password. When the node receives both values, it finds the relevant auth line for your username, splits the rest of the line into the salt and hash, then hashes your password with the salt and checks the results against the configuration value.

Then (importantly) it throws away the password because it knows who you are.

> ### Salt is Healthy in Moderation
>
> An alternate authentication approach lets you add username/password pairs to your *dogecoin.conf* file directly. Beware of this; it's much less secure than the approach described here. If someone were to read the contents of your configuration file, they would be able to read your passwords.
>
> With this salted approach, you still want to keep your configuration file safe, but an attacker will have much more difficulty figuring out your password.

If your password were directly available in memory or on disk, it'd be available to attackers. If the server always used the same salt, then an attacker could try a bunch of potential passwords with the same salt to find something that lets them in. By using a different random salt for every user, attackers have to do a lot more work.

As well, changing your password (and salt) every now and then is healthy, as long as you don't change it so frequently that you have to keep writing it on your hand.

Understand the Risks

You can do a lot of interesting things with RPC commands and a node that has a wallet connected, but anyone who can authenticate and send RPC commands to your node can potentially do things with and to your wallet. Unless you *really* need a full wallet connected to your node, you're safer running in -disablewallet mode.

If you *do* need a wallet, use layers of security such as a good firewall, binding your node only to trusted network interfaces, et cetera.

Please also note that there's no reason you have to run the Python code on the same computer as where you're running your node. It's safer if you don't. That way your password and the salted, hashed password aren't on the node at all, so they're not both available to attackers.

If you need to store your password somewhere, for example with an automated process, be sure to store it securely in a way that it's not also exposed to attackers. This is the weakest part of the entire security model, so if you're going to keep it in a file on your server in the cloud somewhere, realize what you've exposed yourself to.

  # Tip #25   Restrict Node Network Access

Whether you run the Core daemon dogecoind or the Qt GUI dogecoin-qt (see Understand Core Programs, pp. 31), if you allow RPC access, you give up some degree of security and isolation in return for getting data from and potentially changing data in your Core node. This is a tradeoff. It's often worth it, *if* you manage the security implications carefully.

Requiring authentication to access your node's RPC server is essential (see Authenticate RPC Securely, pp. 73), but it's not the only step. You can restrict which machines can access your RPC node in multiple ways.

Do-Nothing Security

By default, a Core node allows RPC requests only from the machine it's running on. If you do nothing, set no configuration, make no changes, your node will take requests only from processes running on your own machine. This is pretty good security.

This only applies if you're running the Dogecoin Core software and if there are no bugs or exploits in your machine or network that give anyone else access to your system. Any other Dogecoin software you should inspect the code yourself (or ask someone you trust to do so) and verify the results. For example, on a Linux machine you could run the netstat utility to list all of the programs listening for network connections:

```
$ netstat -tpl | grep dogecoin
(Not all processes could be identified, non-owned process info
 will not be shown, you would have to be root to see it all.)
tcp    0   0 localhost:22555     0.0.0.0:*  ... dogecoin-qt
tcp    0   0 0.0.0.0:22556       0.0.0.0:*  ... dogecoin-qt
tcp6   0   0 ip6-localhost:22555 [::]:*     ... dogecoin-qt
tcp6   0   0 [::]:22556          [::]:*     ... dogecoin-qt
```

These specific options tell netstat to show all TCP sockets open for listening and the names of the programs listening on those sockets. You can see that the Qt GUI is running and listening on both the IPv4 localhost and IPv6 ip6-localhost interfaces for port 22555. Port 22556 is listening on all interfaces (0.0.0.0 for IPv4 and [::] for IPv6).

Port 22555 is the default RPC port, so this is exactly what we want to see. Remember, however, that a malicious binary *could* fork off of the Qt process, bind to a different port, and change its process name so it won't appear in the list, so look at the *full* netstat output (it's truncated here) before you consider yourself completely safe.

In Dogecoin Core 1.14.6 and earlier releases, you can start the program with the configuration option debug=http to ask the Core to log information about the addresses it binds to for RPC commands. This will add information to your file *debug.log* in your default Dogecoin directory. It can create a *lot* of output, depending how busy your node is, so it may be worth using only while you're testing things. Alternately, your author filed Dogecoin Core issue 3216[3] to discuss improving this situation.

Binding to Specific Interfaces and/or Hosts and/or Ports

Of course, if you *want* to run your Core node on a machine on a network *and* access RPC from somewhere else, you have to make some security tradeoffs.

For example, if you have configured a host running in the cloud to allow IPv6 access everywhere but IPv4 access only from your trusted network, you could use the rpcbind configuration option to restrict the Core to listen to only a single IPv4 address. Add something like this to your *dogecoin.conf* file or script you use to launch the server:

```
rpcbind=10.0.0.3
```

Assuming you have a private network setup where this is the correct IP address of the host running the Core *and* you want to forbid connections from everywhere else, *and* you want to allow only IPv4 connections, this will work.

Add multiple rpcbind options if to listen on multiple network addresses. Mixing and matching IPv4 and IPv6 works here as well. When it becomes too complicated to manage a long list of listen addresses, or if you have other hosts on that network you don't trust fully, you can use rpcallowip to tell the Core to accept only traffic from IP addresses which match the configuration. For example, if you trust only one machine:

---

[3] See https://github.com/dogecoin/dogecoin/issues/3216, fixed in https://github.com/dogecoin/dogecoin/pull/3217.

```
rpcbind=10.0.0.3
rpcallowip=10.0.0.4
```

You can also use netmasks or CIDR notation to specify a range of machines. Be cautious doing so unless you're certain you control which machines can and cannot somehow gain these IP addresses on your network.

Finally, if you want to add obscurity to your network (or you have routing concerns that make certain ports unreliable or unusable), use the `rpcport` option to ask the Core to listen on a port other than 22555[4].

```
rpcbind=10.0.0.3
rpcallowip=10.0.0.4
rpcport=22554
```

Now anyone probing at your network for an open Dogecoin Core RPC port will have to do a very small extra amount of work to figure out what's going on. This is not sufficient security on its own, but it can enhance your security if you've already taken other steps.

Understand the Risks

Restricting network access is important and good. By default, if you make no configuration changes, your node will have good security, at the expense of usability. Consider how much access you need and whether there are other ways to make this information available.

If you do make changes to restrict network access, test and verify them thoroughly, not just by examining listening ports but by testing access to your node. Use a program such as `curl`, `nc`, `dogecoin-cli`, or any other utility.

Remember also that configuration options *change* default behavior. If you assume that your configuration is always set in a *dogecoin.conf* file and will never be modified, re-check whenever something interesting changes, such as re-launching your node. If an attacker gains access to this file, they could change your policy altogether and make your day worse.

---

[4]While it's tempting to use the Stonecutter's secret emergency number, 912 is less than 1024, so you need `root` privileges. Don't run your Core as root.

Tip #26   Enhance RPC Calls

Because Dogecoin has Bitcoin, Litecoin, and Luckycoin in its pedigree–and in its code–you can often find guides and documentation written for those coins that work for Dogecoin.

Sometimes you can't, however.

For example, the current major version of Bitcoin (version 24.x) supports several RPC calls that use and manipulate and query address labels. The current major version of Dogecoin (version 1.14.9) doesn't support these features.

You're not out of luck if you want these features. You just have to be a little creative.

Proxy RPC Calls

Think of a label as a nickname. Any operation you want to do on an address, you can do on a label instead. If you have the ability to associate a label with an address, you're off to the races.

What do you need to make this work? Well you *could* install a toolkit named `Finance::Dogecoin::Utils` from CPAN[5], as it provides a command named `dogeutils`, or you could do the work yourself.

What do you need?

Associate Labels with Addresses

First, you need a way to associate a label with an address. Bitcoin has an RPC call named `setlabel` which does this in the current wallet. It takes two arguments, the `address` to associate and the `label` to use.

This already suggests a lot of the implementation.

The `Finance::Dogecoin::Utils` library uses this information to populate a JSON file inside a configuration directory (*~/.local/share/dogeutils/* on a Linux system). This JSON file contains an object which maps a label to an address like:

```
{
    "Dogecoin Book Tips":
        "DAY5wNkebzEyqUXCkN9koKNBuzXRKRTjcL"
}
```

---

[5]Use the command `cpanm Finance::Dogecoin::Utils` if you have Perl installed. Otherwise see https://platform.activestate.com/chromatic/Dogecoin-Utils for Windows, Mac OS X, and Linux installers.

Writing this as a plain text file has several benefits: you can move it between machines and you can edit by hand if you want, and it's not tied to any wallet format. You can read or write it with anything that understands JSON.

The drawback is that it's not attached to your wallet itself, although that may be a benefit. If someone gets their hands on this file, they'll know you have interest in these addresses (especially if your labels are meaningful), but they won't get your wallet itself or your keys.

With the `dogeutils` command provided by the Perl code mentioned earlier, you could achieve this by using the command line:

```
$ dogeutils setlabel DAY5wNkebzEyqUXCkN9koKNBuzXRKRTjcL \
    'Dogecoin Book Tips'
```

Proxy RPC Calls to a Node

Next you need a way to make calls to a running Core node. First, set up authentication (see Authenticate RPC Securely, pp. 73). With a username and password, you can use any HTTP client or library such as `curl` or Python's `requests` to make requests of the node and relay back the responses.

Handling passwords securely can be tricky. You *could* use an authcookie solution. The Perl library uses a username and password approach[6] with credentials stored in a JSON file in the configuration directory.

When the `dogeutils` program starts, it looks for a username provided on the command line (or in an environment variable) and then looks up the password for that username in the authentication file. If found, any call it makes to the actual Core node will send that username and password.

With authentication set up, now you need to call a method via HTTP. In the example of Bitcoin's `setlabel` call, you must make a POST requests to your node's IP address on the right port to a URL containing your username and password, providing a JSON body containing the appropriate data.

With a Core running on your machine, connect to `localhost:22555` with a URL like `http://user:pass@localhost:22555/` to POST encoded JSON with a Content-Type header of `application/json`:

```
{
  "jsonrpc": "1.0",
  "id": "some identifier here",
  "method": "setlabel",
```

---

[6]Until someone suggests something better!

```
  "params": [
    "DAY5wNkebzEyqUXCkN9koKNBuzXRKRTjcL",
    "Dogecoin Book Tips"
  ]
}
```

Note how close that looks to the `dogeutils` command earlier.

Wrap Calls That Don't Exist

What good is this?

Besides `setlabel`, Bitcoin provides an RPC call called `getreceivedbylabel`, which returns the amount of coins received by the address(es) associated with a label. This does the same thing as `getreceivedbyaddress` except its argument is a label, not an address.

This is fixable. `dogeutils` uses a little bit of glue code to provide both `setlabel` (which doesn't talk to a Core node) and `getreceivedbylabel` (which does). The *implementation* of the `getreceivedbylabel` behavior is entirely within the `dogeutils` program. Given a label, it looks up the associated address in the JSON file, then calls `getreceivedbyaddress` with the address and returns the resulting JSON verbatim.

That's it.

As long as you have some way of communicating with a Core node, some way of managing user authentication, and some type of local storage (hard-coded data, a JSON file, an SQLite database, whatever), you have lots of options.

Understand the Risks

RPC against a Dogecoin Core node needs authentication. Maybe you grow weary reading this reminder. Enable authentication and manage it securely.

The authentication approach described here has a flaw; leaving your password in the clear even in a file you control means anyone who can get access to that file has your password. It'd be more secure to store it in a local keychain somewhere. Alternately, you could force the secure entry of your password on the command line every time you run a command like `dogeutils`.

This is probably secure enough if you're running this program against a Core node running on the same machine, because that's just as vulnerable to anyone who can modify your *dogecoin.conf* file and add password entries or read the authentication cookie. If you're running your utility over the network (or connecting to a Core over a network), ensure that the network is secure.

What Can You Do With This?

You can use this to provide RPC calls not yet supported by the current Dogecoin Core, but you can provide other features as well.

Want to get all of the transactions at a specific block height? Chain together several RPC commands and make your own utility call.

Want to list blocks in reverse order? You can do that.

Want to turn addresses into labels in the output? You can do that too.

You're also not limited to *modifying* calls. You could provide an access control mechanism where certain user accounts can access specific types of calls. For example, you might have read-only accounts that can read general blockchain data such as looking at blocks or transactions and provide other accounts that can create transactions or work with wallets.

Tip #27   Take Actions on New Blocks

Dogecoin is an ever-increasing network of transactions. Each transaction occurs as a specific item in a block. Each block has a specific order in the blockchain, and each block comes in at a specific time. If you think of Dogecoin that way, you can see it as a network of transactions that's also a series of events.

This is a powerful idea you can use to do many interesting things. Consider payment processing. When a block gets mined, look for any payments to any address you're interested in, then do something based on the source address, the destination, the amount, any script in the transaction, or whatever.

You could also monitor the health of the network as a whole by looking at time between blocks, number of transactions in a block, number of coins transferred in a block, difficulty change in blocks over time, or any other piece of data available.

Dogecoin Core gives you options to treat these events as events, so you can do these interesting things and more.

Configuring blocknotify

Dogecoin inherited a Bitcoin feature called blocknotify. This is a configuration option available from the command line or set in *dogecoin.conf* which allows you to ask the Core to launch an external command whenever the Core processes a new block.

You can pass two options to this command: the *number* or *height* of the block and the *hash* of the block. With that, you have everything you need to do much, much more.

How do you make this work? First, configure your node for RPC. Be aware of the risks of doing so, and follow all the security guidelines to your degree of comfort and safety. Second, write a command that does something useful. Third, launch or re-launch your node.

That's it.

Processing a New Block

Let's start by writing a simple command that shows basic block statistics, such as the number of transactions, difficulty, time, and size of a block. This code re-uses some example code to perform authenticated RPC (see Enhance RPC Calls, pp. 80), so you can focus on the behavior:

```perl
#!/usr/bin/env perl

use v5.038;

use JSON;
use Path::Tiny;
use RPCAgent;

exit main( @ARGV );

sub main( $height, $hash ) {
    my $config = get_config( $ENV{CONFIG_FILE} );
    my $rpc    = create_rpc( $config );
    my $block  = $rpc->get_block_by_hash( $hash );

    my $stats  = analyze_block( $block );

    say <<~END_HERE;
    Found block $height
    Processed at $stats->{time}
    Contains $stats->{num_tx} transactions
    Size of $stats->{size}
    Difficulty of $stats->{difficulty}
    END_HERE

    return 0;
}

sub analyze_block( $block ) {
    my $num_tx = $block->{tx}->@*;
    my $time   = localtime $block->{time};

    return {
        num_tx     => $num_tx,
```

```
        time       => $time,
        difficulty => $block->{difficulty},
        size       => $block->{size},
    };
}

sub get_config( $config_file ) { ... }
sub create_rpc( $config )      { ... }
```

If you haven't read much Perl, don't fret at some of the syntactic details. The first handful of lines set the version of Perl (the latest stable release, as of this writing, to use a few newer features) and load a couple of useful modules, including RPCAgent, the secure authentication code. All *that* code does is wrap calls to the Core's RPC listener in a Perlish interface, so that the line $rpc->get_block_-by_hash($hash) doesn't have to manage the details of providing the right HTTP headers. Everything's already set up.

The interesting work is in main() and analyze_block().

main() starts the action. It reads a configuration file (more on that in a moment) containing data to set up an object to perform RPC requests.

When this code runs, it gets two arguments, the height and hash of the new block. The RPC call gets all of the block's data with that hash. The Core returns that data as a JSON data structure, but the RPCAgent code turns that into a Perl data structure (a nested hash, or a dictionary as you might call it in Python, or a hashmap in Java, or an object in JavaScript, or...).

From there, analyze_block() extracts useful data. Difficulty and size are obvious fields. The block's time is in a field named time, but that records seconds since the epoch, so the code uses Perl's localtime to turn that into a textual representation. Finally, the *number* of transactions is interesting, so the code uses the standard Perl idiom to access the items in the tx field (representing transactions) as a scalar value, representing the count of items in the array.

But I Don't Read Perl!

"Context? Scalar? That's unique! Also, wow, look at the sigils!" *De gustibus non est disputandum*, but your author *also* wrote a book called *Modern Perl* that explains all this stuff and more, so head over to http://modernperlbooks.com/ and download a free copy. You'll have one more interesting tool in your toolkit!

When `main()` gets the results, it prints the transaction's description. It will look something like:

```
Found block 4485057
Processed at Wed Nov 23 15:15:28 2022
Contains 12 transactions
Size of 3979
Difficulty of 12696685.226509
```

To use this code yourself, you'll need to set up the configuration file and create an `RPCAgent` object. If you have much programming experience, you can probably imagine what they look like anyhow.

Launching Your Program

How does this get executed? Start by adding a line to your *dogecoin.conf* file to add this command. This shell wrapper sets up the execution environment correctly[7]:

```
#!/bin/bash

cd "${HOME}/dogecoin-tricks-book/"
export CONFIG_FILE="./chapter_3/env.json"

perl -Ilib bin/show_block_stats.pl $*
```

This file sets the path to the RPC credentials configuration file and makes sure Perl knows where to find the `RPCAgent` module. All that's left is to add a single line to *dogecoin.conf*[8]:

```
blocknotify=/bin/bash \
  "${BOOK_HOME}/chapter_3/bin/launch_listener.sh" \
  "%i" "%s" >> blocks.log
```

If you've set up a `cron` job, this will also look familiar.

The `%i` parameter is the height of the new block and the `%s` parameter is the hash of the new block. Whenever the Core processes a new block, it'll launch a system process and pass those two values as variables. In this case, the system will invoke the `bash` shell which will itself invoke `perl`.

As with other changes to *dogecoin.conf*, you'll have to restart your node before the Core will start executing this command.

---

[7] As you might do for a cron command.

[8] Linebreaks are for formatting; they're not needed.

Understand the Risks

Any time you use RPC against a local Core, you should enable authentication and authorization (see Authenticate RPC Securely, pp. 73) and keep unauthorized activity away from your node. This is especially true if you have a wallet attached to your node. Anyone who can connect to your node via RPC can get a lot of data you might not want to expose and take network actions you might not want to permit.

Second, be aware that you could write buggy code[9]. The abstraction here of a shell script calling another program lets you test your RPC code in isolation by calling it with block heights and hashes you already have on hand. That leaves you free to test the *dogecoin.conf* integration later, rather than having to do both at once–especially as any change to the *dogecoin.conf* code means relaunching your node.

If you do get your program into a crash loop, you might harm the stability of the system and the experience for you and others on the machine. Don't let that stop you from doing interesting things but do be careful to think about what could go wrong.

What Can You Do With This?

Perhaps you can light up a lava lamp whenever you make or receive a transaction.

You could update a counter on a webpage or post a message to Slack or Discord (see Post to Discord, pp. 64) with every new block mined.

You could plot the size and fullness of blocks over time.

You could keep a list of the busiest wallets over the past day, week, or month.

What interesting ideas do *you* have now? Be creative!

  Tip #28   Add an Action Launcher

When you developing things that require changing your *dogecoin.conf* file, you'll inevitably run into situations where you need to restart your Core node to test your changes.

Waiting for a node to restart can be annoying on a fast local machine, where you have to wait several seconds to validate all the blocks you have downloaded. On

---

[9]Your author's first few attempts had bugs!

a remote machine in the cloud with a full node, you could wait *minutes* or more between restarts.

We can do better.

One approach is to add features to the Core that let you configure its parameters through RPC calls. Another approach is what software developers happily refer to as "yet another layer of indirection". The latter has at least one really good opportunity to streamline your life.

Launch a Launcher for New Block Actions

The tip for Take Actions on New Blocks, pp. 83 showed off a Dogecoin feature to launch an arbitrary program for every new block accepted on the network. Whether that's one program or a thousand you want to launch, it's all the same to the Core; it's just a single configuration line in the *dogecoin.conf* file.

The contents of this file are completely arbitrary. You could do something, but you could also do entirely nothing. You could do a dozen things. Anything you can do could change. There's only one change you *can't* make without restarting your Core node: the command to invoke.

The previous tip had a configuration like this:

```
blocknotify=/bin/bash \
  "${BOOK_HOME}/chapter_3/bin/launch_listener.sh" \
  "%i" "%s" >> blocks.log
```

... but if it changed to something like this:

```
blocknotify=/bin/bash \
  "${BOOK_HOME}/chapter_3/bin/launch_loader.sh "%i" "%s"
```

... then the file `launch_loader.sh` might be:

```
#!/bin/bash

HEIGHT=$1
HASH=$2

# launch the first block listener
cd "${HOME}/dogecoin-tricks-book/"
export CONFIG_FILE="./chapter_3/env.json"

perl -Ilib bin/show_block_stats.pl $HEIGHT $HASH \
    >> show_block_stats.log

cd
```

```
# launch the second block listener
bash bin/show-desktop-notifications $HEIGHT \
    >> show_desktop_notifications.log

# launch the third block listener
ruby bin/some_cool_code.rb $HASH \
    >> some_cool_code.log

# post something to a webhook
curl -X POST ...
```

In plain English, all this file has to do is prepare the arguments the Dogecoin Core provides (height of the newly-mined block and its hash), do whatever environment manipulation is necessary, and launch *one or more* commands.

Write this launcher in whatever language you like: shell, Python, PowerShell, Rust, or whatever you like. If you want to launch all of these processes independently in parallel, you can. If you want to launch them one after another, you can.

Best of all, you can make changes to the loader without restarting your node. Test this by providing your own arguments to it; it's just a program, launched by the Core automatically but a program runnable however you want.

Understand the Risks

There's only one risk of this technique not already addressed in other examples: anyone who can edit this file can make your Core node launch programs.

If you put authentication information for Core RPC calls in this launcher, they're available to other programs. If you have authentication information for other systems or services, such as a database or a cloud machine or a web service, they may be available to other programs launched from this launcher too.

It's not easy to predict every way someone might find to interfere with your system, but keep this file locked down as tightly as you do your Core configuration.

  Tip #29　Add Desktop Notifications

Now that you know how to do things when the Dogecoin network accepts new blocks (see Take Actions on New Blocks, pp. 83), it's time to get creative. You could go the Internet of Things route and turn on a lava lamp, play a trumpet salute through the office speakers, or flash a message on the electronic reader board outside your school[10].

Let's start with something simple: pop up a desktop notification on Linux, Windows, or Mac OS X with interesting data about each new block.

A Simple Desktop Notification for Block Difficulty

Like other Bitcoin-related cryptocurrencies, Dogecoin adjusts the difficulty of mining a block based on the average time it takes to mine a block. If more miners have more power on the network, the difficulty will increase. If fewer miners are running (they have solar power and it's cloudy), the difficulty will decrease. This variable difficulty is an attempt to keep the amount of time it takes to mine a block to about a minute no matter how much or little mining capacity is available.

If you're a miner, this is interesting because mining difficulty and hashrate affects the likelihood that you will mine any single block.

If you're not a miner, it's an interesting number to track to gauge what's going on in the network. It's also easy but not trivial to get from a Core node, which makes it a good example.

Think about how this will work. In simple terms:

- **when** a new block is mined

- **get** the difficulty of that block

- **display** a message with interesting information

The first part is easy. Now comes some code.

Display Latest Block Difficulty Notifications

What does this code need to do? The Core will launch it every time a new block gets mined. It needs to receive a single argument: the height of that block.

Use the `getdifficulty` RPC call to get the difficulty of the new block. This call takes no arguments and returns the current difficulty. That's also easy enough; it's an RPC call like anything else. All that's left is popping up a desktop notification.

---

[10]Please get permission before doing any of these things!

90

Figure 4.1: Dogecoin Difficulty Desktop Notification

On Linux, the `notify-send` command program (from `libnotify`) handles all of the details. You can provide it a title for the notification, text for the notification, and even an icon. The result looks like Dogecoin Difficulty Desktop Notification, pp. 91. The code can be:

```
#!/bin/bash

HEIGHT=$1
DIFFICULTY=$(dogecoin-cli getdifficulty)
DIFFICULTY_FORMATTED=$(
    numfmt --grouping ${DIFFICULTY_RAW}
)

notify-send \
    -i "$HOME/.icons/dogecoin.png" \
    "Dogecoin Difficulty Update" \
    "Dogecoin difficulty is now "
    "${DIFFICULTY_FORMATTED} at block ${HEIGHT}"
```

That's not so bad, but there's a little bit going on behind the scenes.

The `HEIGHT` assignment gives a readable name to the one parameter this script gets, the height of the freshly-mined block.

The `DIFFICULTY_RAW` assignment uses the `dogecoin-cli` utility to make the `getdifficulty` RPC call. Using the CLI program directly does two things: avoid having to deal with authentication for a call on the local machine and get the result directly as a plain number, rather than JSON output needing additional parsing.

The `DIFFICULTY_FORMATTED` assignment uses the GNU `numfmt` utility to turn a big number into something formatted more appropriately for display. In an English-speaking locale like `en_US`, that means adding commas to separate thousands. Big numbers are easier to read this way.

91

Finally, the call to `notify-send` takes three arguments. First, the `-i` argument uses a path to an icon. This example copies the Dogecoin logo from the core (*src/qt/res/icons/dogecoin.png*) into a separate directory. The second argument is the title of the notification. The third interpolates the nicely-formatted block difficulty and the block height into the notification body.

That's it.

This message will pop up every minute or so and gradually fade away after a couple of seconds.

Understand the Risks

What can go wrong here?

You don't have to have a full node running locally, or even a node with a wallet. You can have a slim node or a headless node or a wallet-free node, but the code as written requires a node running locally.

To connect to a node on another machine securely, be sure to set up authentication appropriately (see Authenticate RPC Securely, pp. 73) and provide the appropriate connection arguments to the dogecoin-cli call.

If you install third-party code (such as the notify-send port on Windows or coreutils on Mac OS X), beware that the code you install is what you intend to install, that you get it from a reputable place, and that anything you download has risks of exposing your information.

If you use this code to affect a third-party machine or system (such as playing the chorus of Queen's "Another One Bites the Dust" on every block mined), be aware that your coworkers, roommates, fellow commuters, and everyone around you may not fully appreciate your musical taste[11]. Be kind to others even if they don't share your excitement about block mining.

What Can You Do With This?

Monitoring difficulty may not be super interesting, but reifying actions on the chain into events untethered to the network can do a lot of things.

You could update stats to display on your web page.

You could estimate time between payments settling if you take Doge for in-person transactions or e-commerce transactions.

You could bring additional mining capacity online if you see the difficulty rate or hashrate drop.

You can leave this as it is and bask in the warm glow that you're part of a community that processes transactions in big blocks, day and night, every sixty seconds or so.

  Tip #30   Generate Transaction Receipts

Dogecoin is money, and money implies a lot of things. Sending and receiving money is all well and good, but besides letting your wallet keep track of your transactions, eventually you'll want to figure out where your Dogecoin came from, where it went, and why and when–not just how much you have left.

At some point, you're going to need to take your transactions to your tax preparer, your accountant, or the spreadsheet you use to track your finances. Fortunately,

---

[11] Sure, it's not "Radio Gaga", but it's also not "Crazy Little Thing Called Doge"!

getting this data out of the Dogecoin Core in a format you want can be straight-forward.

List Transactions

The listtransactions RPC command in the Dogecoin Core allows you to get transactions from your wallet. It provides an swath of data for each transaction, including whether you sent or received money, the timestamp of the transaction, the amount of the transaction, transaction fees, the number of confirmations of the transaction's block, and any label associated with the address.

By default, listtransaction shows only the 10 most recent transactions, but you can provide a second argument to it to show, for example, up to 93[12]. The first argument, while optional, should always be "*", for historical reasons.

```
$ dogecoin-cli listtransactions "*" 93
[
  { ... }
]
```

If you run this command against your wallet, you'll get a list of up to 93 trans-actions, with lots of data. You might not want to send all of this to your receipt printer or accountant. You might want to filter this data more, perhaps by looking only at transactions where you've *received* Dogecoin.

Filter Transactions

The jq program is a command-line JSON processor which allows you to manipulate JSON data, such as the output of most dogecoin-cli/RPC commands. If you pipe the output of the previous command into jq, you can see a nicely-formatted (and probably colorized) view of the data, but there's a lot more power lurking here:

```
$ dogecoin-cli listtransactions "*" 93 | jq .
[
  { ... }
]
```

jq also allows you to filter the data based on the value of a specific field. To filter only transactions where you've received payments, the .category field must contain the string receive. To do this, use a jq command that looks at every element of the JSON array (. []), pipes the results to a filter (|), and then selects only those

---

[12]If you have fewer transactions, you'll get fewer results.

94

array elements where the value of the .category field equals the text receive. It's more work to describe the result than to write that command:

```
$ dogecoin-cli listtransactions "*" 93 |
  jq '.[] | select(.category == "receive")'
[
  { ... }
]
```

Format Specific Transaction Data

That's better, but it's still a lot of extraneous data to give to your tax preparer[13]. What fields of the results do you really want to provide? Definitely amount and transaction time, but probably also the label associated with the address, assuming you're diligent about tracking things like "consulting invoice #1234 for Cool Client" or "Steve should finally pay back that $20 I loaned him in 2014".

jq lets you select specific fields from the JSON data. Add another filter and describe what looks like another JSON object:

```
$ dogecoin-cli listtransactions "*" 93 |
  jq -c '.[] | select(.category == "receive") |
      {timestamp: .timestamp, label: .label, amount: .amount}'
[
  { timestamp: ..., label: ..., amount: ... }
]
```

Note now the use of the -c option to jq, which produces compact output—one object per line. This is easier to read for humans, now that there's a lot less data on each line. Of course, the Unix epoch-style timestamp might be a little too Unix-nerdy for your accountant[14], so it might be better to convert that value to a human-readable format with jq's strftime function:

```
$ dogecoin-cli listtransactions "*" 93 |
  jq -c '.[] | select(.category == "receive") |
      {timestamp: (.timestamp | strftime("%Y-%m-%d %H:%M:%S")),
      label: .label, amount: .amount}'
[
  { timestamp: "2023-04-13 09:08 UTC", label: ..., amount: ... }
]
```

---

[13] Does your tax preparer ask you every year "Did you make any crypto transactions this year?" Your author has had this experience more than once.

[14] If not, you have found a very nerdy accountant, and that's cool!

That's one more filter in jq, only for the .timestamp field. The String Format Time function lets you describe how you want the timestamp to look. Here it focuses on the Year (capital Y means all four digits), the month, and the day followed by Hour, Minutes, and Seconds.

What Can You Do With This?

You don't have to use jq to process the output of listtransactions. You can use anything you want. Remember though that the power of command line tools, including dogecoin-cli and jq, is that you can combine them to do things that might otherwise take a lot of work. All of the other RPC commands in this book–supported by the Dogecoin Core now and in the future–are amenable to similar types of manipulation and processing. If you exercise your skills working with these tools in these ways, you can find yourself solving problems almost as fast as you can think of them.

## CHAPTER 5

Inside the Dogecoin Core

Dogecoin is a lot of things, and one is the Dogecoin Core wallet. There is no such thing as an "official" wallet or "official" implementation of Dogecoin. Dogecoin is a protocol and a network and the consensus of everyone who participates in producing, consuming, relaying, and mining transactions and blocks. Anyone can write or modify or run their own client.

Yet the Dogecoin Core was the first client, and it remains one of the most popular. Anyone can read its code, modify it, suggest changes, and even verify that the code as published is the same code that builds the binaries. This is good, in that it's tested, observed, analyzed, and deployed widely. It's risky, in that Dogecoin belongs to everyone who participates, regardless of their intentions.

Fortunately, the phrase "anyone can contribute" is true–and it requires less skill, time, and effort than you might think. Your experience and insights and suggestions are valuable to Dogecoin and the Dogecoin Core.

Tip #31 Follow Core Development

How do you manage a global network where the success or failure of your personal financial transactions relies on the work of countless strangers, where anyone can participate in the network even if you don't know who they are or if their intentions differ from your own?

That's tricky! You have to trust that everyone plays by the same rules. To do that, you have to verify that *you* understand the rules and that someone's not trying to sneak something past you.

In the world of Dogecoin and cryptocurrency, this means that all participants must:

- follow rigorous, tested, and well-understood mathematical principles and formulas

- adhere to well-defined rules and parameters

- trust that software used implements all of these things correctly

That's one of the reasons the Dogecoin Core is so important. The source code is available, development and organization occurs in the open, and the releases built from it are done so transparently.

Where is the Core?

At the time of this writing, the source code is available on a site called GitHub, at https://github.com/dogecoin/dogecoin[1].

This website and the `dogecoin/dogecoin` project in particular allow people to cooperate top develop the Dogecoin Core by submitting bugs, asking for new features, improving documentation, translating text into multiple languages, implementing new features, and having design discussions. This is a good place to report a potential bug or discuss a new idea.

> **Productive Discussions and Contributions**
>
> While anyone *can* participate, please be respectful of the time and effort of countless other people. Creating an issue saying "Devs please make price go up!!!" won't help anyone, because Core development deliberately resists any activity that could influence Dogecoin's price.

Navigating the Core

The main project page can be intimidating. It has so many features! It can also be underwhelming. Where's all the activity? To navigate all of this, you need to know a little bit about Git and GitHub.

Git is a software tool that lets developers manage the source code: the instructions on what the Dogecoin Core does and how it does it. All of this source code is stored in a Git *repository*–a reliable archive of the history of changes to that code over time.

The GitHub website helps multiple developers share their Git repositories with each other. While Git itself is a distributed system that allows people to collaborate, it doesn't inherently enforce any "official" prime repository. GitHub does;

---

[1] Though it's unlikely this will change any time soon, if you're reading this in 2099 or even 2029, check to see if this is still true before downloading anything.

only those people allowed to make changes in the `dogecoin/dogecoin` repository can change the code that makes up the software that you're running.

Figure 5.1: Dogecoin Core GitHub

However, anyone can *access* GitHub and create (or *fork*) their own repository to make their own changes. Then they can request other people *pull* those changes into any repository, whether their own or the `dogecoin/dogecoin` repository itself.

That forking metaphor is really important.

If you think of everyone's individual repository as a fork off of the main repository, you may imagine a tree or a river. This holds true in the main repository itself, but rather than forks it holds *branches*.

In the same way that a fork represents something that's different from the main repository, so branches can be different.

In the same way that a fork can produce a pull request (to merge changes back into the main repository), so can branches.

Dogecoin Development Process

If you only ever looked at the main branch (called `master`), you'd think not much ever happened. That's because, until release 1.14.8, all development happened on branches. After the release of 1.14.8 in August 2024, all development happens on `master`, with branches created to prepare for a release with a specific version number.

99

Look at Dogecoin Core Development Branch 1.14.7, pp. 100. A widget reading `1.14.7-dev` switches the web view to show a branch other than the main development branch. This particular version number names a branch that represents "the code that would be released as version 1.14.7".

You can switch that view yourself to see what *was* released in previous versions and what *will be* released in future versions. As of this writing, there are two future versions in active development, 1.15.0 and 1.21, with a potential for a 1.14.9. However, all development now happens on `master`, so you can focus your attention there.

Figure 5.2: Dogecoin Core Development Branch 1.14.7

This image also shows links for Issues and Pull Requests. The former is bugs or feature requests and the latter is code that's in progress. Pull requests always point to a specific branch, so that the developers can keep track of what they intend to merge where and when (ready for development, proposed for 1.21 but not 1.15.0, et cetera).

At any point, you can use the Code link, or the list of files and directories, to search through the code as it existed at any point in time (or, in the case of pull requests, code as it would exist if the pull request were merged to a branch). This is powerful stuff!

What Can You Do With This?

You can learn and do a lot with GitHub and the code, even if you don't consider yourself a developer. Feel free to click around, read issues, think about discussions, and look at pull requests. Contribution is open to anyone willing to invest time and effort to help their fellow shibes.

As always, be respectful, understand the rules, and be kind to others. This tool is essential to the Core developers, so please consider their time and resources.

Tip #32 Verify Core Releases

Several tips emphasize the importance of checking and double-checking your assumptions. With any software you use, you should be able to assure yourself that the code you're running is what you expect it to be. Nowhere is this more true than with the wallet software you're running–or relying on others to run.

At some point you still have to trust that the developers are doing what they say they're doing and the code they've designed and written and tested and reviewed does what it should. Trustworthy developers and maintainers will give you multiple, cooperating mechanisms to increase your confidence and decrease your risk.

When a new Dogecoin Core release comes out, you can verify that the code you're thinking of running has been reviewed, tested, verified, and pristine from undesired modifications.

Release Signatures and Checksums

With the current Core development process (see Follow Core Development, pp. 97) at the time of this writing, Dogecoin Core releases have a Git tag and accompanying release notes. For example, see the 1.14.6 release notes at https: //github.com/dogecoin/dogecoin/releases/tag/v1.14.6.

These notes include three data points essential to verification. First, take note of the person who made the release (Dogecoin Core 1.14.6 Release Notes with Author Highlighted, pp. 101). In this case, it's `patricklodder`.

Figure 5.3: Dogecoin Core 1.14.6 Release Notes with Author Highlighted

What Would Patrick Sign?

Now that you know Patrick performed this release, you can verify another piece of data. Navigate to the v1.14.6 tag, then the *dogecoin/contrib/gitian-keys/* directory. You'll see a file called *patricklodder-key.pgp* (Dogecoin Core 1.14.6 Release Gitian Keys, pp. 102).

Figure 5.4: Dogecoin Core 1.14.6 Release Gitian Keys

PGP? GPG? Make up your mind!

PGP stands for Pretty Good Privacy. It's an specification of asymmetric cryptography, codified for the Internet at large in RFC 4880[a]. GPG stands for GNU Privacy Guard, a free software implementation of RFC 4880. While technically GPG signatures are PGP signatures, you'll hear people refer to them interchangeably.

Install GPG from https://gnupg.org/.

---

[a]See https://www.ietf.org/rfc/rfc4880.txt.

This file contains the public half of a GPG keypair (see Create Asymmetric Keys, pp. 17). Patrick has the other private half. Download this file, then import it into your local gpg keyring:

```
$ gpg --import ~/Downloads/patricklodder-key.pgp
gpg: key 2D3A345B98D0DC1F: public key "Patrick Lodder..." imported
gpg: Total number processed: 1
gpg:               imported: 1
```

When you do this, you give gpg the ability to verify that Patrick did or did not actually sign a file with his private key. He uses his private key to add a signature to a file, then distributes his public key. Look at that *SHA256SUMS.asc* file from the GitHub release. Download it and you'll see[2]:

```
-----BEGIN PGP SIGNED MESSAGE-----
Hash: SHA256

87419c...   dogecoin-1.14.6-aarch64-linux-gnu.tar.gz
d0b7f5...   dogecoin-1.14.6-arm-linux-gnueabihf.tar.gz
3e60c4...   dogecoin-1.14.6-i686-pc-linux-gnu.tar.gz
fc2a85...   dogecoin-1.14.6-osx-signed.dmg
bf6123...   dogecoin-1.14.6-osx-unsigned.dmg
c3dd01...   dogecoin-1.14.6-win32-setup-unsigned.exe
c919fd...   dogecoin-1.14.6-win32.zip
888429...   dogecoin-1.14.6-win64-setup-unsigned.exe
709490...   dogecoin-1.14.6-win64.zip
fe9c9c...   dogecoin-1.14.6-x86_64-linux-gnu.tar.gz
-----BEGIN PGP SIGNATURE-----

iQEzBAEBCAAdFiEE3G7OqL+fGx5N4e5SLToOW5jQ3B8FAmLYV6OACgkQLToOW5jQ
3B+PZgf/fgOBO1ZTLO7Kb6HGLDzN0S9M7BmF4igBPO/9/kd06RobbbOb2b/hzu0O
wo5IWha6XzbzIJ89hAzZiCuYdGPg84hacQzKxdN11hOAKQZH9sjEPng/uPcC0Gug
nE5dJzc7/gDi5Esgbod5cgWSATKNeGRlRnb5nUauimyPMnzr/uDHJkCz4IRsA2Oe
KD2OGTKIuyKY6H2Ex3TALRprBkFfnciZVgOMZZxFP/yH07SjVmF6yeBdNMmgbwv1
YgX4sJyNMjvIJvSWTpJrZszsG5jph5xtRl5Mwz9qcYJQ6CvQqmgu+UXUnTKVkf1A
5OEJ8p4n8j51+K8CZ2DrwEFhy/eS1w==
=cbzO
-----END PGP SIGNATURE-----
```

Patrick generated the PGP signature when he produced the release. If anyone has tampered with this file since then, the PGP signature will be invalid. Validate the signature of the file with:

```
$ gpg --verify ~/Downloads/SHA256SUMS.asc
gpg: Signature made Wed 20 Jul 2022 12:29:49 PM PDT
gpg:                using RSA key DC6EF4A8BF9F1B1E4DE1EE522D...
gpg: Good signature from "Patrick Lodder <...>" [unknown]
```

The signature is good; the file is pristine its generation and signing. If you haven't verified Patrick's key yourself, you may get a warning that GPG cannot prove that the signature belongs to him. You can ignore this warning for now[3].

---

[2]The signatures here are truncated to fit the book; they're much longer in the actual file.

[3]The solution is to verify Patrick's identity in person and import his signature directly from him.

What Did Patrick Intend?

What do you know now?

- Patrick made this release

- Patrick created a file containing signatures for all of the files in the release

- Patrick signed the signature file with his private GPG key

- The signature file hasn't changed since Patrick generated and signed it

This can give you a lot of confidence that the signatures of the files listed in the signature file are trustworthy, *if you trust Patrick and the other Core developers*[4].

With all of that established, download the file corresponding to your operating system. Then run a command like:

```
$ sha256sum dogecoin-1.14.6-x86_64-linux-gnu.tar.gz
fe9c9c...   dogecoin-1.14.6-x86_64-linux-gnu.tar.gz
```

If the *full* signature for the command you ran on the file matches the signature from the file Patrick signed, you can have confidence that the file you downloaded is the file Patrick and the Core developers wanted you to download.

Who Watches Patrick?

You don't have to take only Patrick's word for it though. For every Core release, the developers create a GitHub issue to track the release tasks. For the 1.14.6 release, you can see this at GH issue 2975[5]. On this ticket, multiple developers (maintainers and contributors alike) independently build the release files on all platforms and provide the signatures and all of the signatures must match before the release can go public (see Reproduce Builds, pp. 121).

Look at the comments on the issue; Patrick, Michi, chromatic, xanimo, and Alam all performed this task. Everyone's signatures matched in the comments and in the signatures file. Any interested developer can perform this validation for a work in progress, and if their signatures do not match, the Core maintainers can stop the release process to investigate what's going on.

---

[4] You don't have to trust the Core developers. Your author thinks they're trustworthy, but your author also will not tell you how to set your own trust and risk levels.

[5] See https://github.com/dogecoin/dogecoin/issues/2975.

Understand the Risks

Whew! That's a lot of work to ensure that the files provided are the files the developers intended to release. For someone to tamper with this, they'd have to go to a lot of effort to compromise a developer's keys, replace a lot of data in a lot of places, or generate a malicious file that somehow produced exactly the checksum of the original file.

The latter is only *theoretically* infeasible, to the best of everyone's knowledge now. If at some point in the future this becomes practical, the Core will have to switch to another verification/validation mechanism. Whenever this happens, keep your eyes and ears open for lots of publicity about how to ensure that what you're running is actually what you intend to run.

What Can You Do With This?

Hopefully all of this process improves your confidence that the Dogecoin Core you're running is the Core the developers created and verified and released. This work doesn't prevent bugs or misfeatures, of course, but it helps guard against imposters and frauds sneaking unwanted code onto your system.

If you get your Core from a third-party repository, such as a Linux distribution vendor, Homebrew, Chocolatey, or another packaging system, review their process and pipeline to see if they perform these verifications. If they don't, tread with caution. The Core developers can't control what third-party bundlers, packagers, or redistributors do—so you can always verify and then download the files the Core developers produced.

  Tip #33   Build Core in Docker

One of the benefits[6] of free and open source software like the Dogecoin Core is that anyone can inspect, build, modify, and redistribute the code itself, provided they follow the license and can actually download and compile software.

Anyone can do the former. The barrier to entry for the latter is a little bit higher. Thanks to good tools and practices, it's within the reach of people willing to put in a little bit of work installing and configuring software.

---

[6]Also one of the frustrations.

Docker for Dogecoin

A program called Docker has two important features you can use here. First, it lets you run containers, which are stripped-down instances of operating systems that are independent, self-contained, and isolated from the rest of your system. Second, it lets you run Linux in containers whether you're running Windows, Mac OS X, or Linux.

Getting Docker in Place

If you're using a package manager that comes with your Linux distribution or something like `homebrew` on Mac OS X, you can probably run `apt install docker` or `brew install docker` and follow the instructions. Windows is a little bit more work.

If you can't use a package manager like this, `docker.com` has instructions[7]. As dozens of tutorials available from your favorite search engine will attest, you need to download and install a Docker client and server and may need to make a couple of changes to your operating system to allow it to switch between running your system as it is and running one or more containers.

Because this is external software you're installing and configuring, be cautious that you're downloading from a reputable source. Cross-check against more than one article and guide to make sure the recommendation for which source and site to use is trustworthy[8].

Building a Container to Build Dogecoin Core

When you're done, you should have the ability to run a *Dockerfile*. This is a series of instructions for the `docker` program that tell it how to build an image from which you can launch one or more instances of a container. If you're not tired of the analogy that a text file is a blueprint for software, it's like that[9].

To build the Dogecoin Core, you need a *Dockerfile* that can:

- Bundle a Linux distribution

- Provide enough Linux tools to get the Dogecoin Core source code

---

[7] At the time of this writing, https://docs.docker.com/desktop/ has more details.

[8] For example, Microsoft's most current guide also points at Docker Desktop; see https://learn.microsoft.com/en-us/virtualization/windowscontainers/quick-start/set-up-environment?tabs=dockerce.

[9] If you *are* tired of that analogy, feel free to open an issue with this book and suggest an improvement.

- Provide the tools required to build the Dogecoin Core

A three-bullet list make this seems straightforward. The *Dockerfile* makes it re-peatable and, better yet, easy enough to modify once you know what you're doing. Create a file containing:

```
FROM ubuntu:22.10
WORKDIR /build

# depends systems prereqs
RUN apt update \
 && DEBIAN_FRONTEND=non-interactive \
    apt install --no-install-recommends -y \
        autoconf automake binutils-gold  ca-certificates \
        curl build-essential faketime git libtool pkg-config \
        python3 bison python3 python3-pip python3-wheel \
        python3-setuptools autotools-dev less libssl-dev \
        libevent-dev bsdmainutils libboost-system-dev \
        libboost-filesystem-dev libboost-chrono-dev \
        libboost-program-options-dev libboost-test-dev \
        libboost-thread-dev libqt5gui5 libqt5core5a \
        libqt5dbus5 qttools5-dev qttools5-dev-tools \
        libprotobuf-dev protobuf-compiler libqrencode-dev \
        libdb5.3++-dev libdb5.3++ libdb5.3-dev

RUN python3 -m pip install setuptools --upgrade

RUN git clone https://github.com/dogecoin/dogecoin.git \
    && cd dogecoin
```

This seems like a lot, but it's four commands. First, build a Docker image using the Ubuntu Linux distribution, version released in the year 20**22** and the **10**th month (October).

Next, use the Linux command apt[10] to update the system and install a bunch of packages, including everything you need to download, configure, and compile the Dogecoin Core.

Third, ask the Python language to update a library it uses to install other libraries. This will save a little bit of time and trouble.

Finally, check out the Dogecoin Core source code.

---

[10]It's a command from a slightly different Linux system called Debian, to which Ubuntu owes a huge debt. Credit where it's due!

Save this *Dockerfile* in a directory by itself, then build the image. On a Linux or Mac OS X system or from a PowerShell terminal on Windows[11], the command will be something like:

```
$ docker build -t dogecoin-build-image -f Dockerfile .
```

First, by putting this file in its own directory, the act of building the image will go a lot faster. If you put this file in your home directory, Docker will have to look at every file in that directory and every file in every directory recursively under that directory before it can finish building the image. You'll notice it's slow.

Second, every unique line in the *Dockerfile* gets built once and only once. If you change a line, Docker will rebuild that step and everything that follows in the file. By arranging the lines this way, it's easy to add to the file and rebuild the image hopefully without making your life more difficult.

Building Dogecoin Core in Docker

Run the command `docker run -it dogecoin-build-image:latest`. Here `dogecoin-build-image` is the name you provided when you built this image. This is a tag. `:latest` is a discriminator used to distinguish between different *versions* of the images you've built[12].

This command says "Run the fresh image, named `dogecoin-build-image`, and let me interact with it from the command line.". If all goes well, you'll see a shell inside the container of an Ubuntu Linux system, within a directory named */build/dogecoin*. Finally you can configure and build the core itself:

```
$ ./configure && make && make check
```

That command will run for a while and will eventually complete (if all goes well) with Linux binaries available in *src/* and *src/qt/*. If anything goes poorly, you can report bugs to the Dogecoin GitHub repository (see Open an Issue, pp. 118).

Understand the Risks

Downloading and compiling software from the Internet always carries risks. Even though the Dogecoin Core has a strict and stringent development process intended to reduce the risk of contamination or security problems, there's no guarantee that breaches won't happen.

---

[11] Your author hasn't actually tried this; feedback welcome.

[12] You don't have to remember this. It may help you in the future.

Building the Core in a container provides some degree of isolation from the other code and data on your system. This is no 100% guarantee of safety; there may be bugs or exploits in Docker that let malware escape the sandbox.

Of course, you face similar risks anytime you *run* code downloaded from the Internet.

What Can You Do With This?

The more you learn about Docker and container management, the more options you have. This technique is helps Core developers test builds on multiple operating systems and versions of operating systems.

With a reusable development environment like this, you can track Core work (see Follow Core Development, pp. 97) and test proposed changes, make changes and see how they affect the code, even try fixing bugs. There's no requirement you do any of this; you could merely spend some time, learn a few things, and enjoy yourself playing around.

On the other hand, you may find it satisfying to have the ability to build the code you're running yourself, rather than relying on (and trusting, see Verify Core Releases, pp. 101) other people to build it for you correctly.

Maybe you're running on a platform for which the Core developers provide no binaries, perhaps a specific operating system or machine type. Maybe you have other versions of libraries you want to use, or the default configuration isn't to your liking. With the *Dockerfile* provided, you have the ability to launch a new container any time you want and try whatever you want and those containers won't interfere with each other. When you exit the container, you can save your work and restart from there.

This can be very handy; you can launch a new container and immediately check out a different GitHub branch or pull request or make your own branch. Or, if you find yourself working from the same branch all the time, you could add another RUN line to the *Dockerfile* to build a new image.

  Tip #34   Put Your Face on Your Wallet

Dogecoin intended to put an amusing face on cryptocurrency from the start. That's one reason a friendly Shiba Inu shines on you happily when you launch the core GUI and look at the wallet. A happy dog makes other people happy.

Dogecoin also has the value of decentralization. The source code is available for anyone to inspect, modify, fork, or redistribute. The network is open for anyone to use any well-behaved client, even if it has nothing to do with the Core. As the core developers do more and more work, network operators have more and more options to choose how their nodes and the network as a whole behaves.

Some of these options are available to everyone running the latest software. Other options rely on your ability to reconfigure the Core (or write your own software). Some of these are terrible ideas, but others can be harmless fun, including replacing the default wallet image with one of your choosing. For example, see Your Author As a Post-Apocalyptic Troubadour Shibe, Hosting a Wallet, pp. 110).

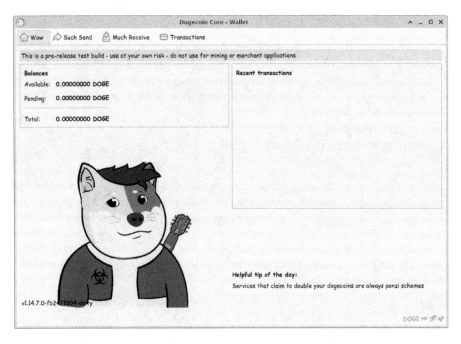

Figure 5.5: Your Author As a Post-Apocalyptic Troubadour Shibe, Hosting a Wallet

Assuming you can compile source code (see Build Core in Docker, pp. 105), apply a patch, and supply an image with the appropriate characteristics, you can brighten your own heart too!

Replacing an Image in Dogecoin Core

While this may seem like either a lot of work or a little bit, your author's first proof of concept took two lines of code. Doing this in the mostly-right way might take about ten lines of code. With the right guidance, swapping an image can be straightforward. Some caveats apply.

How Images Work in Dogecoin

The Dogecoin Core GUI uses a graphical toolkit called Qt[13] to mange interface widgets, handle events, and display graphics. Because `dogecoin-qt` (see Understand Core Programs, pp. 31) is an all-in-one single binary file, it has to include image data within the one binary file rather than load images, icons, and graphics from disk whenever it starts[14]

This means that replacing an image isn't as easy as moving, renaming, or overwriting a file.

Finding the Right Place to Modify

Dogecoin uses a Qt feature called *forms* to implement the visible user interface. Every screen and view you see when you use the Core GUI has a form definition somewhere. These forms specify the types of widgets, the layout of the screens, and the images, fonts, and other resources used to draw the interface. Find these forms in *src/qt/forms/* in the Dogecoin source code tree.

Similarly, all of the images used in forms are in the *src/qt/res/icons/* directory. To find a specific image, look at the files in that directory. Spoiler alert: the one used in this tip is *src/qt/res/icons/wallet_bgcoin.png*. Remember that name.

Working with full filenames and paths is verbose, so Qt in Dogecoin uses a `QResource` file to associate short aliases with these filenames. If you look at the file *src/qt/bitcoin.qrc*, you'll find that the wallet background image has an alias of `wallet_bgcoin`. In retrospect, that was predictable.

The file *src/qt/forms/overviewpage.ui* contains the code to modify. Find it by searching the Core for the alias name.

You can go one of two directions here. Either you've used Qt this way before and you know exactly what this means, or you take two more steps and figure out what's going on. Behind the scenes, when you compile Dogecoin Core, the Qt software uses the XML in this UI file to generate complex but predictable C++ code from it. This C++ code has a specific class with public members you can manipulate like you would any other C++ code. Build the Core to see a header file named *src/qt/forms/ui_overviewpage.h*, which you can read to see that the

---

[13] See https://qt.io/.

[14] This is a good thing; there are risks to loading external images.

corresponding class is `Ui_OverviewPage`. The class member corresponding to this image's widget is `label_wallet_bgcoin`.

> **You're Not Born Knowing This**
>
> Knowing this now is all well and good, but if you didn't know this, how would you figure it out? Developer tip: get good at searching a directory tree for a string of text. Your author had to find the right file, then searched for the filename and found the `qrc` file. If you're using an IDE which includes search features, use them! Otherwise practice using a search utility like `grep`, `git-grep`, `ack`, or `hg`. You'll unlock a superpower figuring out how things fit together.

If you're comfortable modifying C++ code, you're off to the races now. The last thing you need to know is that the Core initializes this form in a class called `OverviewPage`, found in *src/qt/overviewpage.cpp*. To load a *different* image, either modify the Qt resource and UI declaration, or change the pixmap[15] when the Core initializes this page.

Hard-Coding a New Image in the Core

Initialization occurs in the `OverviewPage` constructor. This constructor receives the Qt widget as a parameter called `ui`, and all of that widget's public members are available. To override the wallet background image, create a new `QPixmap` and set it as the pixmap of `ui->label_wallet_bgcoin`:

```
QPixmap pic("src/qt/res/icons/jr_dev_puppet_by_real_shibes.png");
ui->label_wallet_bgcoin->setPixmap(pic);
```

Recompile and launch `dogecoin-qt`. You should see the results from Your Author As a Post-Apocalyptic Troubadour Shibe, Hosting a Wallet, pp. 110.

What Kind of Images Work?

The original image is a 300x300 square image in PNG format with a transparent background. You can use any image you like, but the closer you stick to this format, the better your results. For your best options, stick with a square image of about those dimensions, in PNG format with a transparent background. JPEG

---

[15]This is how Qt refers to an image in this context.

files don't allow transparency and GIFs don't necessarily have the color fidelity you might want.

Working with the Existing Patch

Hard-coding an image file and recompiling every time you want to re-skin your wallet background is more effort than you might want to undertake. What if there were a way to select an appropriate image every time you *launch* the Qt GUI or use a configuration option to provide a path to an image?

As of this writing, you can't do this with any released Dogecoin Core version, but you *can* grab a patch from the author's GitHub branch[16] and compile the code yourself. This ten-line change adds the command-line option `--walletimage` which, when provided, will attempt to load a file at that location (either an absolute path or relative to the directory where you launched the Core GUI) to use as a background image.

For example, to use that Junior Developer Puppet image created by Jimmie[17] of the Real Shibes[18] podcast[19], your author would use a command like:

```
$ ls *png
jr-developer-shibe.png

$ dogecoin-qt --walletimage=jr-developer-shibe.png
```

Understand the Risks

This tip deliberately includes *no* walkthrough of how to apply or build this change because of the risks it introduces. You're using *unsupported, modified* software that has even fewer guarantees than the Dogecoin Core, which is released with zero guarantees. If something goes wrong, you're own your own.

Beyond that standard disclaimer, the risk is higher than running a modified build. If you use an arbitrary image outside of your own control, you put yourself at risk of malicious code or exploits from an image. Yes, from an image–bugs exist in image processing libraries, and multiple security vulnerabilities have happened because people assumed image files are safe.

Using an image you created, modified, edited, and validated with your own software on a machine you trust is one thing. You're still using code completely unvet-

---

[16] See https://github.com/chromatic/dogecoin/tree/put-your-face-in-dogecoin.

[17] https://x.com/jimjimmiejames

[18] https://x.com/Real_Shibes

[19] Listen and subscribe at https://www.youtube.com/channel/UCua0kZDo5fa8srcWfxlmA5A.

ted by the Core developers, but the provenance of that image is less worrisome. If, instead, someone tells you "Download this file, then run your core wallet with this command-line option to use this image!" never do it. Fun customizations entirely under your control are one thing. Letting someone else put arbitrary code or data on your computer is something else. Avoid those people.

Furthermore, if anyone points you at a branch, a patch, or any other modification to the Dogecoin Core and tells you to download, compile, and run it, be *very* careful. There's no end to the malice someone could perform if they convince you to do something you wouldn't normally do.

If you're an experienced developer who followed the explanation here, you have enough information to evaluate that the changes do what they should do, but that's no substitute for the entire Dogecoin Core development community evaluating a proposed change and finding and fixing bugs and misfeatures. Even something simple like the base of the branch could have serious implications for the performance, security, and stability of your system.

If, after all of these disclaimers, you want to continue experimenting, follow the standard rules:

- Know the risks

- Back up your wallet and seed phrase(s), and test new code and changes with an empty wallet

- Read and think about the code carefully

- Double-check all sources and references

Above all, have fun! Yes, that's a long list of caution to keep in mind, but if you're careful, even silly ideas like this have value! Writing software can be enjoyable and it can teach people to think deeply about all the ways things can go wrong even as they're still putting friendly faces and interfaces on their systems.

Tip #35   Understand the Debug Log

If you launch the Core–especially if you don't launch the Qt GUI–a lot of things happen, many invisible. You can go on your way, leaving your VM running in the cloud, laptop plugged in on your desk, or tiny solar-powered device nestled in a corner of your goat barn, and your Core node will contentedly hum along. While you're enjoying the sunshine, your node is making and receiving connections to relay data, validate transactions, and gather new blocks.

Unless you've set up your node to do something visible whenever something happens (see Take Actions on New Blocks, pp. 83), you'll have to look at Dogecoin's debug log to get more details about what's going on silently. Let's dig in!

Where is the Log?

By default, the Core writes debugging information to a file called *debug.log* in your data directory. If you've configured your own data directory with the option `datadir`, look there. Otherwise look in a directory named *.dogecoin* under your home directory on a Linux, BSD, or other Unix-like system. Look in the *Library/Application Support/Dogecoin* directory under your home directory in a MacOS system. Look in a directory *%APPDATA%\Dogecoin* on a Windows system.

In all cases, you should see a text file named *debug.log* that slowly grows as your node continues to run.

For Your Console Only

If you really need to debug something–if you're *changing* the core, for example–use the configuration option `printtoconsole` to bypass writing to *debug.log*. This is only useful if you're *launching* the Core from the command line. Remember that output will go away when the terminal does unless you redirect or capture it.

What's in the Log?

The Core prints a lot of messages to the long as it starts up. Most of these are internal details and not very interesting, but in the case where you're running on a slow machine (perhaps a tiny device, an underpowered cloud VM, an aging laptop), you can watch the log file to see the progress.

Some of the initial data can be valuable for your on purposes. For example, you might see something like this snippet:

```
2023-04-15 21:26:05 Dogecoin version v1.14.7.0-cb9a47611
2023-04-15 21:26:05 Default data directory /home/user/.dogecoin
2023-04-15 21:26:05 Using data directory /home/user/.dogecoin
2023-04-15 21:26:05 Set backupdir "/home/user/.dogecoin/backups"
2023-04-15 21:26:05 Using backup directory
                    /home/user/.dogecoin/backups
2023-04-15 21:26:05 Using config file
                    /home/user/.dogecoin/dogecoin.conf
```

```
2023-04-15 21:26:10 Bound to [::]:22556
2023-04-15 21:26:10 Bound to 0.0.0.0:22556
```

You can see a few important details, including:

- the version number of the Core[20]

- the default and actual location of the data directory

- the directory where the Core will store wallet backups

- the location of the configuration file used

The last two lines show that this node is listening on every available network interface for IPv4 and IPv6 traffic. If you're concerned about specific network security, use this information to help figure out if you've configured your node appropriately.

What Does It Mean?

After your node has run for a while, you'll see other types of messages:

```
2023-04-15 21:27:49 receive version message: /Shibetoshi:1.14.6/:
    version 70015, blocks=4678120, us=67.189.98.209:42198, peer=3
2023-04-15 21:32:19 connect() to [2602:fc05::32]:22556 failed
    after select(): No route to host (113)
2023-04-15 21:36:09 UpdateTip: new best=af8461i... height=4678121
    version=0x00620104 log2_work=75.419327 tx=95260772
    date='2023-04-15 21:34:26' progress=1.000000
    cache=78.0MiB(10374tx)
```

This example shows three different types of information, all useful. In your log, these three lines will all be on single lines, but to fit printed pages, they've broken across multiple lines.

The first line records a new connection from another Dogecoin node. This node reports that it's running version 1.14.6 of the Core with a standard version (see Set Your Node Comment, pp. 38). It uses version 70015 of the Dogecoin protocol (the standard current version), has about 4.7 million blocks available (remember this number), has an IPv4 address and port, and is the third peer connected to the running node.

---

[20]Looks like your author was using a modified development version, probably something useful for the book.

The second line indicates a network connection failure to another node over IPv6. This is normal; nodes come and go. Sometimes backhoes make entire networks of nodes temporarily invisible.

The third line will approximately once every minute. That probably gives away what's happening. This node has received a new block with the given hash (truncated in this example), at height 4,678,121. There are now 95,260,772 transactions in the entire blockchain, and the block's mining date was April 15, 2023 around 9:30 pm UTC. This block is the most recent block (otherwise `progress` would be less than 1.0 or 100%).

The `log2_work` number represents the number of block hashes you'd have to calculate to recreate the entire chain up to this block: two to the power of 75.419327 in this case. It measures the effort that's gone in to calculating the chain of blocks of which the current block is the latest. You don't have to understand all of this here.

If your node has been offline for a while and needs to sync, you'll see a lot of `UpdateTip` messages. If your node is fully synced, you'll see log lines as new blocks get mined.

What Can You Do With This?

If you're the author of this book, asking "How does one know how many transactions are in this system" (see Calculate Your Dogecoin Footprint, pp. 250), you can look at that third log line and think "Wait a minute, *something* is counting that already" and realize you have much less work to do than you thought[21].

If you're interested in mapping nodes on the network, you can analyze node connection logs to see how many nodes are available, the versions of the network they support, the versions of the Core they run, their node comments, and any other information you can derive from things like IP addresses and block counts.

If you're interested in mining statistics, you can examine the `UpdateTip` messages to determine the approximate mining difficulty at any point, the number of seconds between blocks, the average number of transactions per block at any period of time, et cetera. While this information is available historically from other sources and commands, taking it from log data is cheap and easy for people who are comfortable processing logs as they arrive.

---

[21] Sigh?

Tip #36　Open an Issue

No matter how smart, hard-working, and attentive software developers are, sometimes things just aren't quite right. You'll find a bug, or something that looks like a bug. You'll look for a feature you're convinced really ought to be there and it isn't. The documentation will be confusing, or missing, or full of glaring typos[22].

In short, you're going to notice something that could be improved. In that case, who better to ask for the improvement than you?

How to Open an Issue

As of this writing, Dogecoin Core development happens on GitHub (Follow Core Development, pp. 97). If you run into a developer in person, follow them on Twitter, swap messages on Reddit, or change your node's comment (Set Your Node Comment, pp. 38) to read "Hey devs, when are you porting Core to the Nintendo Switch?", your request isn't *official*-official until it's captured in some persistent fashion.

That's why the other Core developers aren't facepalming right now, thinking "Wait, is he *really* trying to port to the Switch?"

If you have a more realistic request, such as "I see an error message when I try to ..." or "I think there's a missing feature, because I can't figure out how to ...", then you're in luck. Visit the Dogecoin Core GitHub Issues page[23]. You'll see something like Dogecoin Core GitHub Issues, pp. 119.

Check to see that the URL includes `dogecoin/dogecoin`. Note that the GitHub interface has `Issues` underlined, indicating that you're looking at the right issues tab.

Now you have to make a choice. Are you going to report a bug or request a new feature? Depending on your choice, you'll see a textbox with some text already included. Both options have different templates that will prompt you to include lots of information. Fill out as much detail as you can. While the more detail you can include, the better–but if you don't have everything, it's okay. Think through all that you do know. Developers can always ask for more information.

---

[22]Unlike this book, in which all typos are subtle and intentional.

[23]At https://github.com/dogecoin/dogecoin/issues/.

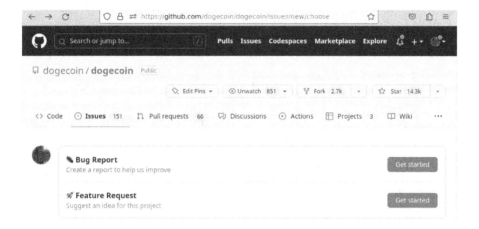

Figure 5.6: Dogecoin Core GitHub Issues

What Makes a Good Request?

Before you submit your request, think through what you're trying to achieve. A good request tells a story. That story needs at least four parts.

- What you hope to accomplish. What's your big goal? It's important to describe this broadly.

- What prevented you from accomplishing that goal. Please be specific. Did you get an error message? What was its exact text? What did you try? What are the steps you tried; can you reproduce them?

- What context you have. What operating system are you using? Processor? Version? The more detail here, the easier it is discover useful information. This is often more helpful for bug reports than feature requests, but sometimes features are missing because of bugs.

- How you will know your problem is solved. This is different from your big goal; this is very specific.

A Practical Feature Request Example

For example, your author filed a feature request in his head while writing about footprint calculations (see Calculate Your Dogecoin Footprint, pp. 250). What would that feature request look like in this framework?

119

I want to accomplish: performing specific calculations based on actual, measurable statistics of the blockchain, such as the number of transactions seen at a specific block, accessed via RPC commands such as `getblockcount`.

What prevented me: there's no obvious RPC method that provides this data directly or implicitly provides this data. While it's possible to data for each block from the genesis block to the most recently-mined block and calculate this count myself, that would require hundreds of thousands of RPC requests and would take a lot of time.

What context is appropriate: A test on a 64-bit Linux machine with a wallet enabled and transactions indexed (Index All Transactions, pp. 192) demonstrated that that the node has enough data to provide this information.

My problem will be solved when: anyone can use a single RPC command to get the information needed in fewer than ten seconds.

This request turned into a pull request[24], where all of this information is also important. Hopefully you can see how the information from the feature request explains how a developer might go about understanding and solving the problem.

What Can You Do With This?

The issue tracker is open to anyone who wants to register on GitHub and open an issue. *Should* you?

If you've found what you believe to be a bug–the Core behaving in a way you don't expect, a feature not working, a strange crash or error message–then *please* do report a bug. Developers spend a lot of time thinking about things, testing them, and trying to improve quality, but the best feedback comes from real users doing real things in new ways and different contexts.

If you're missing a feature, think about the best place to add it. Is it something that should be in Core, or is it something you can add yourself with the features the Core already provides? (In the example case, is it *better* done in Core, because the external implementation would be expensive, difficult, or fragile?)

If you're not sure, feel free to reach out to other Dogecoin friends for advice. Talking through your issue with someone else is a great way to clarify your thoughts and help you answer the template questions well. If it turns out your idea would be better as a discussion ("What if we could put a node on a satellite?" or "Is it worth switching from scrypt to another proof of work algorithm?"), there are other, more

---

[24] See https://github.com/dogecoin/dogecoin/pull/3248.

appropriate forums for that–and that's okay![25]

If you're on the fence, it's probably better to open an issue than not. Feedback is useful, even when it's "I think this isn't quite right". Free and open source software gets better the more people that participate well, and user feedback is an important way to build the right things together.

Tip #37 Reproduce Builds

Before the Dogecoin Core maintainers can release a new version, they have to be certain that the pending release has certain properties. It has to run, it must pass tests, it must be compatible with the Dogecoin network, and it must have certain security properties. It also has to be reproducible.

For anyone to *trust* that the source code available in the release reflects the actual binary releases you can download, anyone must be able to produce the exact same output–bit for bit identical–to what the maintainers have published.

This work is not easy. It's essential to allow users to trust but verify. Otherwise how can you know that someone hasn't sneaked in an unexpected change that puts you and your security at risk?

Fortunately, the maintainers of Dogecoin Core and other cryptography projects have put in a lot of work to answer this question–not just how to verify the integrity of a Core release (see Verify Core Releases, pp. 101), but to ensure that Core releases *can* be verified. You can help!

Gitian and Build Reproduction

The Dogecoin Core maintainers use a tool called Gitian[26] to manage the process of downloading, verifying, and building dependencies, then building the Core itself, for every released platform. The resulting binaries can be compared between Gitian builds (especially by different people on different systems at different times!) to ensure that they are identical (see Make a Hash of Fingerprints, pp. 1).

Gitian can make three errors obvious:

- The build process fails, which means developers need to debug the issue.

---

[25]Look through existing discussions; a lot of these questions have come up before.

[26]See https://gitian.org/.

- The build process succeeds, but the resulting binaries are not identical, which means developers need to fix or eliminate any non-determinism in the process

- Gitian is configured or launched incorrectly, which usually means someone like your author is testing the wrong code. It happens. Keep reading!

Setting Up Gitian

To perform Gitian builds, you need a Linux development system with a lot of disk space (at least 20 GB free), a decent Internet connection, and some time. This could be a laptop, a virtual machine, or a dedicated server. You also need enough system administration skills to install and configure the necessary Gitian dependencies.

The Gitian documentation is good and worth reading. The Dogecoin Core also includes a Gitian setup guide[27]. One benefit of the latter is that Dogecoin developers do their best to keep it up to date to reflect the state of the world and any bugs, quirks, workarounds, or improvements that people have discovered.

Once you have set up Gitian, you can test it.

Watch for the Gitian Bat-Signal

Not all changes require a Gitian test. If you're suggesting an improvement to documentation or translation, that's probably safe to assume the output is deterministic. The Core maintainers will test this anyway when they start to produce a release, but you don't need to worry about it.

If someone upgrades a dependency, such as a library that the Dogecoin Core requires to run, it's essential to test that builds remain reproducible. In that case, the pull request in the Core repository (see Follow Core Development, pp. 97) will get a label called "gitian check needed". This is your sign that you can help!

Get Ready to Go

Check out the Core git repository on your Gitian system. Now look at the pull request. Suppose it's PR 3364[28], shown in Dogecoin Core Pull Request, with the upstream repository and branch highlighted, pp. 123.

---

[27] See https://github.com/dogecoin/dogecoin/blob/master/doc/gitian-building.md.

[28] See https://github.com/dogecoin/dogecoin/pull/3364.

Figure 5.7: Dogecoin Core Pull Request, with the upstream repository and branch highlighted

You need two pieces of information to start your Gitian build for this pull request: the location of the upstream repository and the name of the branch with the proposed changes within that repository. Click on the link highlighted in the image: `https://github.com/edtubbs/dogecoin/tree/1.14.7-dev-fontconfig`. Split this in two parts. The first part is the repository location, Ed Tubbs's fork of the Core repository, `https://github.com/edtubbs/dogecoin/`. The second part is Ed's branch name, here `1.14.7-dev-fontconfig`. Ignore the `tree/` in the middle.

Start the Build

If you've followed the Gitian setup instructions, you know what to do, but it's still a little bit confusing. Instead, it may be easier to use this Bash alias to launch the Gitian build:

```
function buildgitian {
    bash contrib/gitian-build-test.sh -b --docker -u "$1" -c "$2"
    pushd "gitian-output/dogecoin-binaries/$2"
    sha256sum * | sort -k2
}
```

Add this to your *.bashrc* or *.aliases* file, however you prefer (or turn it into a shell script on its own). Then you can use the command from your Dogecoin core checkout directory:

```
$ buildgitian https://github.com/edtubbs/dogecoin/ \
  1.14.7-dev-fontconfig
```

```
...
e0480f...   dogecoin-1.14.7-aarch64-linux-gnu-debug.tar.gz
56950d...   dogecoin-1.14.7-aarch64-linux-gnu.tar.gz
6b0a28...   dogecoin-1.14.7-arm-linux-gnueabihf-debug.tar.gz
ed0880...   dogecoin-1.14.7-arm-linux-gnueabihf.tar.gz
237c41...   dogecoin-1.14.7-i686-pc-linux-gnu-debug.tar.gz
6824b4...   dogecoin-1.14.7-i686-pc-linux-gnu.tar.gz
cdd318...   dogecoin-1.14.7-osx64.tar.gz
65a6ad...   dogecoin-1.14.7-osx-unsigned.dmg
f64af9...   dogecoin-1.14.7-osx-unsigned.tar.gz
6e5172...   dogecoin-1.14.7.tar.gz
1156be...   dogecoin-1.14.7-win32-debug.zip
df80b0...   dogecoin-1.14.7-win32-setup-unsigned.exe
6b568b...   dogecoin-1.14.7-win32.zip
b93147...   dogecoin-1.14.7-win64-debug.zip
e9cf63...   dogecoin-1.14.7-win64-setup-unsigned.exe
a3baac...   dogecoin-1.14.7-win64.zip
c16fd3...   dogecoin-1.14.7-win-unsigned.tar.gz
1ad74b...   dogecoin-1.14.7-x86_64-linux-gnu-debug.tar.gz
```

This output elides part of the signatures for brevity. After the build finishes and you see this output, you can compare your signatures to those of other developers in the comments of the pull request. If all goes well, you've helped prove an important characteristic of the Dogecoin Core with regard to this proposed change. If there's a mismatch, you've proved something else important: that it's time to debug the problem. Either way, you've helped an incredible amount!

What Can You Do With This?

Many of the proposed changes that go into each release of the Dogecoin Core may produce incompatibilities between builds. It's up to the Core maintainers to make sure that none of these changes actually *do* break the repeatability of builds. Anyone capable of and willing to perform these verification steps can help out with development by attempting builds and producing their signatures.

This process is essential for many but not all proposed changes that go into future Dogecoin Core releases. It's even more essential as the Core developers prepare for an imminent release. The Gitian build process is a great way to increase your ability to contribute to the Core. Furthermore, the latter sections of the setup guide show you how to add cryptographic signatures verifying that you've personally built the to-be-released binaries and they're trustworthy.

CHAPTER **6**

Basic (and Advanced) User Stuff

A Dogecoin wallet isn't a pouch full of paper wads and metal lumps. Neither is it a ledger of incoming and outgoing transactions. It *is* a list of numbers and descriptions. Those are public and private keys and any labels you've attached to them.

Even though that sounds deceptively simple (and doesn't translate directly to the contents of your pocket or your bank account), the contents of your wallet are important. They give (or you) access to the Dogecoin you can spend. What you do or don't do with them can make the difference between having Dogecoin and watching someone else spend it.

Not everything you do with Dogecoin is always deeply technical. This should be fun, too! Sometimes it's satisfying to do something difficult or complicated or unpopular just to say you've done it, but it's also satisfying to *avoid* problems or complications or little annoyances that get in your way.

This chapter covers several ideas to make your use of the Core more pleasant, enjoyable, and fun–while working with your wallet safely and securely and protected from potential catastrophe.

  Tip #38 Launch the Core Silently

Think back to the first time you launched the Dogecoin Core on your computer. A friendly dog's face popped up along with some text in a cartoonish font, as if to say "Don't worry. Things are happening, and they're fine. Don't take it all so seriously. We're going to learn things and have fun."

All of that is true. Yet if you've set up your Core to launch when you launch your computer (or if you launch it yourself to run in the background), you might not need that reminder every time.

Here's how to make things quieter.

Minimize Splashing and Windows

This happy little window is called a "splash screen", both in application parlance and in the Dogecoin source code itself. With that fact, you know almost everything you need to make the Core launch itself a little more quietly.

If you run `dogecoin-qt` yourself from a command line, you might have used the `--help` command to show a list of options you can use to change how the Core and Qt GUI behave. In this example, two options are important:

```
$ dogecoin-qt --help
...
UI Options:

  -min
        Start minimized

  -splash
        Show splash screen on startup (default: 1)
```

In other words, if you launch the Core with `--splash=0` and `-min`, you'll see no splash screen. When the program finishes validating the blockchain and loading your wallet, you won't see the main window either. The program will simply run in the background and do its thing until you ask to see it.

If you're using a Unix-like system such as Linux, a BSD, or Mac OS X and launch the program yourself from the command line[1], you can write a shell script, set an alias, or simply type `dogecoin-qt -splash=0 -min` and not much will (visibly) happen—as you want.

If you use the GUI to launch the Core, you'll have to find the launcher and change its properties to pass these command-line options. Consult your operating system or window manager documentation for more details.

Understand the Risks

The biggest risk to changing how you launch the Core is that you might forget it's started and leave it running. Yes, the happy little Doge icon will still be in your system tray, but if you don't see the splash screen or have the main window pop up when everything starts, everything your Core is configured to do could be running without you knowing it.

That list of *everything* includes sending and receiving network traffic, updating your wallet, and responding to RPC requests. Even if you're certain this isn't an

---

[1] You can also do this from Windows, of course.

issue, double- and triple-check before you let anything launch your Core without you knowing exactly when and how it's running. Otherwise your wallet or data may be at greater risk than you intended.

  Tip #39   Manage Multiple Configurations

In Launch the Core Silently, pp. 125 you learned how to make configuration changes to make your use of the Core more pleasant. There are many, many other potential configuration changes. Run this command from the command line to see more:

```
$ dogecoind -help
```

Every option you see there is an option you can use when you launch Dogecoin, whether it's the dogecoind daemon or dogecoin-qt GUI (see Understand Core Programs, pp. 31).

There are lots of ways to manage these options, ranging from "I remember everything and type the whole command every time I want to launch the Core" to "I have shell aliases" to "I put shortcuts in my program launcher". They all of pros and cons.

There's another way to keep things organized. The *dogecoin.conf* file can contain almost every command-line option you see from the command earlier. Because it's written to a file, you don't have to remember the contents every time you want to launch the Core.

This can be useful.

Manage One Conf File

Consider the configuration to launch the Core minimized with no splash screen: dogecoin-qt -splash=0 -min. The corresponding lines in *dogecoin.conf* are:

```
splash=0
min=1
```

If you add these lines to your existing configuration file then launch dogecoin-qt with no other arguments, you'll see no splash screen and the Core will launch minimized. **Beware** that if you *replace* your existing file with these two lines, you might have to reconstruct your file, so if you're trying this for real, *add* these lines if you already have a file, rather than replacing it.

Manage Multiple Conf Files

Now you can launch your Core however you launch it and your preferences are saved in a way you can always use them and always review them. Suppose you want to launch your Core in multiple different ways, though. Perhaps you want to do multiple things simultaneously, such as:

- Test changes on regtest or testnet

- Perform RPC and block actions with a node with no wallet attached

- Keep track of your own transactions in your own wallet with a pruned node

You *could* isolate all of these operations is separate machines. Another way is to have multiple, separate configuration files.

Make a copy of your *dogecoin.conf* in the same directory, with the name *dogecoin-noisy.conf*. That's a silly name, suitable for doing something silly. Edit the file and change splash=0 to splash=1. Just one bit of a change, in every sense. Save the file, then run:

```
$ dogecoin-qt -conf=/path/to/dogecoin-noisy.conf
```

You should see the Core launch, this time *with* the splash screen. Everything else will behave as you have otherwise configured it. That's silly–but instructive.

> ### Sharing Configurations
>
> Unfortunately, as of Dogecoin 1.14.8, the Core has no feature corresponding to Bitcoin's includeconf configuration option. This option allows you to include *another* configuration file from your *dogecoin.conf*, so if you want to have most things the same but only a few different, you have other options.
> Check subsequent releases for this feature!

Now that you have multiple configuration files, you can launch your Core pointing at the desired file for what you intend to do *and* you can modify or multiply your configuration files. Let the complexity expand to your satisfaction!

Understand the Risks

One risk of using one or more *dogecoin.conf* files is that you may forget what's in each file. This risk is greater if other people have access to your system. Suppose you share a server with your little brother, and he has the ability to use the `sudo` utility to grant himself temporary root access. He could modify your configuration file to grant himself RPC access, point your blocks directory to something untrustworthy, or install an action that makes your Bluetooth X-Wing speaker play honking goose noises every time a new block gets mined (see Take Actions on New Blocks, pp. 83, but you'll have to figure out the Bluetooth and goose noises yourself).

Of course, if you share a server with your little brother and he has `sudo` access, he could also borrow your *wallet.dat*, so pick your family connections well.

Another risk of juggling multiple files is forgetting the context in which you intend to do something and use the *wrong* configuration file. Consider pairing this tip with one of the other techniques such as making different program launchers or shell aliases to run different Cores. For example, you could make separate aliases/launchers named `Testnet Dogecoin GUI` or `testnet-dogecoin-qt` to remind you of what you intend to happen when you launch the Core with that specific configuration.

Finally, if you *do* use multiple configurations to switch between networks and full/pruned nodes and wallets, be very careful to manage options such as `datadir` and `backupdir` (especially since the 1.14.6 release). Keeping data separated is essential to keeping your data uncorrupted and safe.

  Tip #40   Practice a New Language

Some people think English is the language of the Internet. It's actually memes, but they don't know it yet. Until then, it's up to all of us to find ways to communicate with other people in ways they understand.

The Dogecoin network communicates in terms of numbers: big and small numbers, all bundled together in blocks and transactions and addresses and Dogecoin and fractions of Dogecoin. We humans bring meaning to those numbers by agreeing on what they mean and presenting them in ways we can understand: labels, QR codes, addresses, secret keys, et cetera.

The Dogecoin Core itself does the same thing. Whether you're a goat farmer in western Canada, a small business investor in eastern Africa, a polar ice researcher in Antarctica, or a student studying European finance in Austria, you balance two important things when you use the Core: what the underlying information *means* and how you *prefer* to consume that information.

In other words, without all this highfalutin' talk, you should be able to read use the Core in any written language you prefer[2]. Good news: more and more people have that option!

Language Setting

Run the `dogecoin-qt` program and ask for its help output; you'll see an option called `--lang`:

```
$ dogecoin-qt --help
...
UI Options:

-lang=<lang>
      Set language, for example "de_DE" (default: system locale)
```

If you've configured your computer to use German spoken in Germany as your default language, the Core should do the right thing *if* the developers have provided the appropriate translations. You should have to do nothing special with Dogecoin for the software to behave normally.

Suppose instead you're about to go on a trip to Portugal and Brazil and want to brush up on your Portuguese, so you'd like to manage your transactions in one or both languages. How do you figure this out?

How Language Codes Work

The example code, `de_DE`, has two parts. The first is `de`, which represents Deutsch, the German language. For more details, see ISO-639-1[3]. The second part is DE, which represents the country of Germany, known as Deutschland to its inhabitants. For more details, see ISO-3166-1[4]. The combination of language and country code means "German, as spoken in Germany". Similarly, `en_US` refers to "English spoken in the United States of America" and `en_GB` refers to "English as attempted by the residents of that island we defeated back in the 18th century"[5].

This is the ideal behavior, anyhow. If there's no specific UK variant of English or Canadian variant of English, it's okay for the translation system to fall back to

---

[2] Your author recognizes the irony of writing this whole book in English.

[3] Wikipedia explains this at https://en.wikipedia.org/wiki/ISO_639-1

[4] Wikipedia also explains this at https://en.wikipedia.org/wiki/ISO_3166-1_alpha-2

[5] "You'll Be Back" not withstanding.

generic English. Similarly, if there's a Portuguese in Portugal language code pt_-
PT but no Brazilian Portuguese pt_BR, it's okay to fall back to Portuguese, even
though you'll probably embarrass yourself with country-specific slang.

How do you know what the Core actually supports? Look at its source code[6]
to find a list of all current translation files. Look for your language code and
any specific country code. If there's a file present, you can use that language.
Otherwise, you'll have to choose something else.

Understand the Risks

The biggest risk you face with this technique is fleeting disappointment. Test this
with the --help command to dogecoin-qt:

```
$ dogecoin-qt --help --lang=af_ZA
Gebruikerkoppelvlakopsies:

...

-lang=<lang>
     Set language, for example "de_DE" (default: system locale)

-resetguisettings
     Alle instellings wat in die grafiese gebruikerkoppelvlak
     gewysig is, terugstel
```

You may find, as in this case, that only some of the text has translations. Alternately, you may have no translations for your language and/or country. Think of
this as an opportunity, however. This could be your chance to contribute to the
Core and help countless other shibes and potential shibes use Dogecoin in their
own preferred languages!

The other risk is that you inadvertently set a default language to something you
don't understand well enough to disable[7]. If this happens to you, examine your
configuration file for the lang setting and change it to your preferred locale setting. Be aware that, at least as of 1.14.9, using the --help command does not
process your configuration file but instead will use your system's default locale
setting.

---

[6]See https://github.com/dogecoin/dogecoin/tree/master/src/qt/locale

[7]Your author does not admit to setting the default language on a fleet of French laser printers to
Bulgarian from the United States on accident on the first of April once.

Tip #41   Swap Wallets

You've decided, for security purposes, to maintain a wallet of watch-only addresses representing addresses you use to send and receive Dogecoin (Watch Wallet Addresses, pp. 157). If you mostly receive and rarely send, this can be a great way to avoid mishaps and keep your hard-earned koinu safe.

Sometimes you do need to spend your coins and send transactions from your addresses. Because your watch-only wallet lacks your private keys, you can't use it for sending transactions (at least, not transactions that you need to validate with private keys).

In the case where you want to use Dogecoin Core to manage your transactions, consider running the core with different wallet files.

Launch with Wallet File Locations

Suppose you have two *wallet.dat* files, one containing private keys and addresses and the other containing watch-only addresses. The hot wallet has a filename of *hotwallet.dat* and the other wallet has a filename of *watchonlywallet.dat*.

> ### Make Multiple Wallet Files
>
> How did you get two different files? Launch the Core. Let it sync. Let it create a wallet (if you don't have one already).
> Close the program and wait for it to exit. Then move or rename the *wallet.dat* in your Dogecoin Core data directory. Launch the Core again. Now you have two wallet files with no connection to each other.

When you launch the Core, use the configuration option `wallet` to specify one or the other wallet file:

```
$ dogecoin-qt -wallet=watchonlywallet.dat
```

When the Core launches, you should see the difference between the different wallets based on their contents. Use a separate configuration file (Manage Multiple Configurations, pp. 127) for different wallets to make this easier

Be aware that all wallet files, whatever their names, must be in your Dogecoin data directory.

Understand the Risks

If you launch the Core with a new wallet but use the same blockchain data, your Core will have to rescan blocks since its last launch with that wallet. This helps prevent data loss (specifically any transactions that affected your wallet), but it will increase startup time. If you need fast startup time, launch the Core with your other wallet(s) frequently–once a day, once a week, however often makes sense for you.

Every computer with a wallet increases your security footprint. Every new wallet file containing private keys adds a security risk. Even if you have encryption enabled on your wallet as you should, keeping a *wallet.dat* of any kind on a laptop, desktop, USB key, cloud server, or anywhere else gives you one more machine or device or location you have to secure.

With that said, running a node that *doesn't* have access to a wallet storing private keys reduces the risk that a bug or an exploit in Dogecoin Core or the network or RPC or REST or any other interface can expose wallet information–so running a node without a wallet (Work Without a Wallet, pp. 151) or with a watch-only wallet can increase your security.

For a *tiny* amount of minimal security against the laziest possible attacker, you could even name your wallet file something entirely *unlike wallet.dat*, just in case anyone accesses your hard drive and searches for files matching *wallet.dat*. This is only a tiny level of protection, but there's no requirement that your wallet have a filename of any specific form.

If you *do* add this additional level of security and obfuscation, you might consider *not* hard-coding wallet names in multiple *dogecoin.conf* configuration files, just to make the files slightly less obvious to any attackers.

  # Tip #42   Limit Logs

To know what's going on with your node, look at a file named *debug.log* in your Dogecoin directory (see Understand the Debug Log, pp. 114). This file contains logs about your startup and running configuration, transaction and network activity, and warnings and errors. Depending on how you've configured your system, this file can grow slowly or rapidly.

By default, unless you intervene somehow, this file can grow large without any bounds. This is probably not what you want, so you can (and should) limit the size of the file, depending on what you need to do and why. While you're more likely to run out of disk space due to the blockchain growing in size, any file that can grow without limits can strain your resources or add extra costs and overhead.

Restart Your Node

The easiest and simplest one-time approach is to shut down your node, truncate or delete this file, then restart your node. The Dogecoin Core will recreate this file when it starts up successfully. If you read this and immediately check and see that your log file is taking up 10 GB of space[8], this intervention can be quick and easy.

There are three downsides. First, you'll lose all information in the existing log, unless you save a copy somewhere else. Second, you'll have to restart your node. That may take a few minutes. Third, you'll have to do this again in the future. Even so, this is a fine approach as a quick intervention.

Limit Log File Size

For a more permanent solution, use the `shrinkdebugfile` configuration option from the command line or in your *dogecoin.conf* file. This parameter takes a boolean true or false value. When true, the Dogecoin Core will truncate your log file when it starts if the file is larger than 10 MB (10 million bytes). Any time you restart your node, the Core will perform this action.

There are three downsides. First, you'll lose all truncated data unless you've saved it. Second, you must restart your node, which may take a few minutes. Third, this works only with the configuration option enabled. If you forget to enable it on the command line or remove it from your configuration file, the Core will contentedly append to your existing log file. These may all be worth the effort.

Rotate Your Logs

The previous two options work well on a desktop machine. On a server, you may want a more robust and standard solution, such as using a technique called *log rotation*. With this approach, a separate process monitors log files on a schedule and moves them out of the way when they meet specific criteria such as age or file size. This rotation generally keeps around the most recent logs and gradually archives and deletes older copies.

On Linux and Unix-like systems, look for a program called `logrotate`. This common program provides all of the options described earlier. For example, you might write a configuration file like:

---

[8]In July 2023, that sounded like a big log file. If you're reading this in July 2033, that may not sound so large.

```
/home/user/.dogecoin/debug.log {
    rotate 7
    daily
    postrotate
        /usr/bin/killall -HUP dogecoind
    endscript
}
```

This configuration applies to the log file *debug.log* in the *.dogecoin* directory belonging to user user. It rotates the log file daily, keeping the most recent seven
entries. After rotating a file each day, it sends a message to the Core to restart,
so the Core can start writing to a new log file. This is a common pattern, but if it
doesn't meet your needs, logrotate has many, many other options to explore.

If you have root or administrative access, move this file into your logrotate
configuration directory, something like */etc/logrotate.d/dogecoin*. If you lack this
access, save this file in one of your own directories (*.dogecoin/* is fine), then enable logrotate in your cron system (perhaps by running crontab -e):

```
18 * * * * /usr/sbin/logrotate /home/user/.dogecoin/logrotate.conf
```

If these commands seem strange or scary to you, that's okay. You don't have to use
them until and unless you're comfortable configuring system services as yourself
or as a root user. There are plenty of tutorials available online to explain in much
more detail.

Understand the Risks

What could go wrong?

Running out of disk space is always a risk, so choosing a strategy to manage the
size of always-growing files is important. The tradeoff to consider is archiving
and eventually deleting old data that might contain something you really wanted
to know. For example, if you fear that an attacker might send your node bad data
or issue deleterious commands you don't anticipate, having logs available (and *not*
on the production system) might help you identify the danger and react to it more
quickly.

A deleted log does you no good. Then again, neither does an unmonitored log.

As with everything else, you have to pick and choose what you're going to learn
about, what you can pay attention to, and the level of risk you can accept.

135

Tip #43   Securely Back Up Your Wallet

While all of the settled Dogecoin transactions in the world are kept in the public ledger (that's the Dogecoin blockchain), your ability to send coins from any of your addresses depends on your ability to access the private keys kept in your wallet. Your wallet is a list of private and public keys, labels for the associated addresses, and a little bit of metadata. That's the most important thing you can keep track of and protect.

The security if your coins is the security of your wallet is the security your private keys. This expresses itself in two ways:

- you need to retain your ability to access those keys

- you need to keep other people from getting those keys

In other words, keep your wallet available and keep it safe. You can achieve the former by making a backup.

Understand the Risks

If you have only a single copy of your wallet, the thing (computer, USB key, laminated paper in a safety deposit box–see Use a Text Wallet, pp. 139) containing that wallet has a physical risk of theft, destruction, or data exfiltration. Hard drives die, cats knock laptops off of bookshelves, and USB keys go through laundry. Maybe your hard drive will grind to a halt. Maybe you'll spill a glass of orange juice on your laptop during a hearty breakfast[9].

Having multiple copies of essential data is a good disaster recovery mechanism. Maybe you're being sufficiently cautious during an upgrade of the Core version. Maybe you want a copy stored on a USB key in a safe deposit box in a bank.

Losing your wallet doesn't mean your coins are gone, but it may mean you no longer have access to them. Making a backup of your wallet gives you the ability to *restore* your access to your keys, which restores your access to your coins.

Creating and securing a backup reduces the risk of losing your single mechanism of accessing your coins, but it increases the risk that other people may gain access to your backup. If you give a USB key with your wallet to your brother in law, the neighbor kid down the street, and your estate planner, then your ability to get the

---

[9]Your author has, of course, *never* done this. It was lunch.

file when you need it is your access problem. Their ability to keep your file secure is now your security problem.

Similarly, if you upload a backup to a cloud provider, their security is your risk and their ability to lock you out of your files is your access problem.

What can you do?

Dogecoin GUI Backups

If you're running the Dogecoin Core GUI with a recent enough release (as of this writing, version 1.14.9), use the File -> Backup Wallet menu item to launch a window that allows you to select the location and name of a wallet backup file.

If you do this, your system will think for a moment, write out the file, then pop up a notification telling you that your backup is complete (see Dogecoin Core Backup GUI Notification, pp. 137).

Figure 6.1: Dogecoin Core Backup GUI Notification

Protect this file. It's the same as the *wallet.dat* file in your Dogecoin configuration directory.

If you open the file in a file editor, you'll see bunch of binary data. This file uses a well-known format called Berkeley Database (BDB). Bitcoin, Litecoin, and Dogecoin have used this format for years[10]. Anyone who can read this file can read its contents. You need something more secure.

Encrypted *wallet.dat* Backups

Back in the GUI, go to the Settings -> Encrypt Wallet menu. If you haven't used this before, it will pop up a dialog that allows you to select a phrase to use to encrypt your wallet.

---

[10]They will likely both move to a format known as SQLite, which is even easier to work with in some ways.

137

A good encryption phrase balances memorability with privacy. Just like with passwords, short and/or easily-guessable phrases are bad (`password`, `1234`, `s3kr1t`). Experts often suggest using memorable quotes then modifying them with typos, substitutions, and enhancements.

For example, use a line from a song, add spice, and throw people off. If you're a John Rox/Gayla Peevey fan, maybe `I want a hippopotamus for Christmas` could remix with Pokey the Penguin to become `I want a skeptopotamus for Christmas 3:-]`, something that's much more difficult for anyone to guess.

With this encryption in place, you'll need to enter the passphrase to do anything with the wallet, so now you have *two* things to keep track of: your wallet itself and your passphrase. Anyone–including you–who has hold of one has to get hold of the other to do anything with either. This is the price and benefit of additional security.

Dogecoin RPC Backups

You can achieve the same outcomes with the CLI/RPC tools. For example, use the `dogecoin-cli` utility to run `backupwallet` RPC command, providing a filename argument. The Core will back up your wallet to a file of that name within the configured backups directory (by default, *.dogecoin/backups/*)[11]). See Dogecoin Core CLI Backups, pp. 138. Similarly, the `encryptwallet` RPC command can add a passphrase to a wallet.

```
wispwheel:~$ dogecoin-cli backupwallet mybackup-2022-12-23
wispwheel:~$ ls ~/.dogecoin/backups/
mybackup-2022-12-23
wispwheel:~$
```

Figure 6.2: Dogecoin Core CLI Backups

Because `dogecoin-cli` is an RPC client, anything you can use to send RPC commands to a Core node can send a `backupwallet` or `encryptwallet` command.

---

[11]This is the behavior as of 1.14.6.

What Can You Do With This?

Where can you store backups that they're safe? Wherever you can easily access the files when you need them and where other people can't.

What reduces your risk that other people can do bad things with your data if they do get your backups? Encryption.

How do you secure your passphrase? Choose something meaningful but not guessable. Make sure it's secure and kept away from your backups. Make it possible for someone you trust to help you restore it in case of an emergency.

Above all, keep a cool head. Think about what could go wrong, and choose your risks appropriately. Make a plan and test it. You'll discover flaws and benefits that allow you to revise your plan before something goes wrong.

  # Tip #44   Use a Text Wallet

Without diving into the complex and fascinating world of semiotics, this book attempts to express the idea that cryptocurrency is the meaning that humans assign to complex sequences of numbers and rules about how to interpret, generate, and verify those numbers. While this book uses the Dogecoin Core as an example of software used for that meaning-making, it is not the only way to participate in the Dogecoin network.

Even the word "wallet" is a metaphor, perhaps a misnomer. If you had 10,000 Dogecoin right now you could make up a valid Dogecoin address, throw away any way to regenerate the key, and send some or all of your funds to that address. The network doesn't care until *someone* attempts to spend those funds. Unlike a real kruggerand which changes hands during transactions, the network only cares about proof of access to any transaction when someone tries to spend the funds.

In other words, you don't need *software* to hold funds or receive funds. You only need software at the point of creating and broadcasting a transaction to *spend* funds.

How do you keep track of how to spend your hard-earned coins without a wallet?

What is a Text Wallet?

The wallet you might have attached to your Core node is a binary file which acts as a database of sorts. It contains a list of private keys, labels, and addresses. Your wallet may also include transaction information to help you access this data more quickly–but the only thing it really must contain is enough private key information to derive public keys and addresses.

A text wallet is *something* which contains that information but which is not connected to a Core node or another piece of software. It could be a text file on a removable drive, a list of data printed on a piece of paper, or even something chiseled into stone. For this to work well, this text wallet must demonstrate at least three properties:

- The data must be accessible and retrievable when you need it

- The data must be secure from theft or loss

- The data must be accurate, complete, and reproducible

As usual, the balance between security (no one should have access to the contents of the wallet) and accessibility (it should be easy to perform wallet transactions when desired) complicates things. If you were to memorize your private keys and could type them from memory perfectly every time you wanted to spend funds, you could avoid having a wallet artifact at all.

You should not trust your memory, however. Too many things could happen to you to rely solely on perfect recall.

Similarly, you could write your private keys on a sticky note and keep it under your keyboard[12]. This would be easy to access and immune to power outages, hard drive crashes, or attackers who manage to gain access to the files on your laptop, but paper is easily lost and damaged, and if someone wanders into your New Years Eve party and sees the note, you could be in trouble.

Generate or Export a Text Wallet

How do you get a text wallet? The rules for *generating* private keys and deriving public keys are well-understood, so any compatible and trustworthy software can work for you. If you want to use the Dogecoin Core, you can start with an empty wallet then expand the wallet pool to give you enough addresses for the foreseeable future (see Replenish Your Address Pool, pp. 160).

If you want to use different software, such as `libdogecoin`, you might find it easier to generate a BIP-32 or BIP-44 wallet (see Derive More Addresses, pp. 318) where you have a single seed phrase which you can use to generate an infinite, predictable series of private and public keys. In that case, you can back up *only* the seed phrase and use it to regenerate your wallet at any time. You may want to generate plenty of addresses in advance, however, to avoid needing to enter your seed phrase too often.

---

[12]Don't do this!

Furthermore, any software which implements either of those two BIPs can generate a master key which you can use to regenerate your wallet. Be cautious that you can trust this code, that it implements the Dogecoin network and key parameters correctly, and that you can trust its entropy source (see Embrace Entropy, pp. 21).

Regardless of whichever approach you choose, you should be able to export a list of addresses to which you can receive Dogecoin as well as either a seed phrase or a list of private keys. While you can keep all of this information together, you may want to separate your addresses from your secrets. This allows you to treat the most secret information with the most security (you only need access when you want to spend funds), while keeping the addresses more accessible (you may need to get a new address or batch of addresses every day).

What Can You Do With This?

To spend your funds, you need to use some kind of software to create a transaction. One of the inputs, of course, is the private key of the address which holds the funds you want to spend. To get that private key, you need to get the right data from your text wallet into the software.

If you've backed up your private keys, you can import the appropriate private key into software like Dogecoin Core. WIF-style keys are good for this (see Interchange Your Wallet Keys, pp. 15). Seed phrases are good for software like (or built upon) `libdogecoin`. Remember that you can import data from a computer file or type it in by hand, but you should check for typos or file corruption before doing anything you can't reverse.

An alternate approach[13] is to use a local copy of the `coinb.in` website with your computer disconnected from every network to create and sign transactions. While this adds friction to the process, it also adds security by reducing the amount of time your private keys or passphrase needs to be exposed.

By separating your addresses from your text wallet, you can monitor them for transactions (see Watch Wallet Addresses, pp. 157) and even perform automated actions on them (see Act on Wallet Transactions, pp. 194).

Understand the Risks

The biggest benefit of a text wallet is that it's not online and not easily available from a computer that someone can carry off or break into. If you keep your wallet on your phone, laptop, or another mobile device, the risk is even higher than if

---

[13]Your author has not tested it, but takes the recommendation of Reddit Dogecoin Shibe /u/Fulvio55 seriously.

you have a clunky desktop computer padlocked to a steel beam in your basement. The more difficult it is for anyone to access your wallet, the fewer chances you have of losing your funds.

The drawback of most effective security measures is that inaccessibility means inconvenience when *you* want to access your funds. It's easy to use a custodial tipbot with a mobile client to send Dogecoin to someone for creating an amazing meme, posting something useful on social media, or being an amazing person, but all of those funds are at higher risk than if you keep your keys printed on paper and stored in a safe deposit box.

Paper has its own risks too. Moths or insects can eat it. Fire and water can damage it. Wars or civil unrest can destroy it (and the bank and the bank vault).

A USB key has other risks–besides physical damage, it can have data loss or you may no longer be able to read the file, depending on its format. (A *wallet.dat* file may be riskier than the plain-text contents of your wallet, for example.) Lightning may strike or an EMP blast may happen if a mad scientist miswires a Tesla coil.

In all cases, *your* ability to remember that the text wallet exists, how to access it, and how to use it is another gating factor. If you've saved up emergency money for health care costs and you cannot go to a physical location or communicate your access mechanism to your loved ones, your funds may not be accessible. The same applies to an unencrypted wallet on your computer too, of course.

In every case, balance the risks of losing your funds with the risks of losing your text wallet. Multiple copies of the text wallet reduce the risk of losing any single copy, but increase the chances that anyone can get their hands on any copy. The safest funds are in a wallet where no one can access the private keys. That probably defeats your purpose though, so your security strategy will depend on your circumstances and needs. Review this regularly and make changes as appropriate.

  Tip #45   Find All Received Addresses

When you first launch a recent release of the Dogecoin Core, it will create a new wallet for you if and only if you've never created one before. By default this wallet will contain a lot of addresses as well–more than you're likely to use for a while.

An address is usually most useful when you can do something with it, either receiving coins or sending them. Extra addresses aren't a problem–there are plenty to go around–but but a wallet full of unused addresses can make accounting a little more difficult than it needs to be. What if there were a way to narrow down the hundreds or thousands of addresses your wallet has generated for you to only those addresses that have actually received coins?

There's an obvious but tedious way, a spreadsheet way, and a programming way.

Manual Transaction Accounting

Click the Transactions button in the Dogecoin GUI. This will open a window showing all of the transactions associated with your wallet. Use the filter widgets to include or exclude types of transactions or date ranges. For example, the figure in Dogecoin Core Transactions List, pp. 143 shows only those addresses which have received funds in transactions.

Examine the addresses (or tooltips, if you use address labels) in that list to see the addresses which have received funds. If you've had this wallet for a while or made a lot of transactions, have fun.

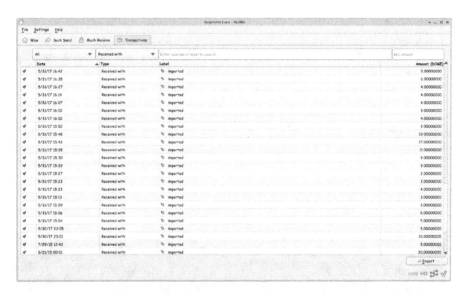

Figure 6.3: Dogecoin Core Transactions List

Spreadsheet Transaction Accounting

Look again at Dogecoin Core Transactions List, pp. 143. The "Export" button in the bottom-right corner can save you time and effort. Click that button and the Core will prompt you to write a comma-separated value (CSV) file containing all of the transactions displayed with your current filter set.

This file contains all of the data shown and more, including a true or false value indicating whether the transaction has been confirmed, the timestamp of the transaction, the type of the transaction (mining, sending, receiving, etc), any label associated with the address, the address, the amount of Doge affected for that address, and the transaction ID.

You can import this CSV file import it into any modern spreadsheet application (Excel, LibreOffice, Numbers, Google Sheets) or manipulate it with just about any programming language you can find and perform all sorts of calculation, charting, or data manipulation you like.

To answer the question "Which addresses have received Doge in transactions", set your filter to "Received with", export the data, grab the contents of the fifth field in the CSV file (starting from 1, not 0), and get a unique list of the contents of that column.

Command-Line Transaction Evaluation

The `dogecoin-cli` program combines well with other other command-line programs in a nerdy Unix way. If you're comfortable using the command line, you can answer a lot of questions quickly and easily.

The `listreceivedbyaddress` RPC command tells the Core to give you a list of all addresses which have received Dogecoin in any transaction. It does the same filtering as the GUI transaction list earlier *and* it groups all transactions together per address[14].

While `dogecoin-cli` often hides the details of *how* the RPC works behind the scenes (JSON over HTTP), complex output is often a JSON object: in this case, a JSON array of objects, as shown in A List of Received Transactions by Address, pp. 144.

```
{
  "address": "                        ",
  "account": "imported",
  "amount": 14.00000000,
  "confirmations": 2760172,
  "label": "imported",
  "txids": [
    "                                        "
  ]
},
```

Figure 6.4: A List of Received Transactions by Address

---

[14] Isn't that convenient, that this command does exactly what this tip set out to do? So suspiciously convenient.

Almost any programming language you're likely to run into these days can work with this JSON output. You don't need to rush off to install PHP or Ruby or Rust for this, though; the jq utility[15] may be all you need.

Provide jq with JSON data and ask it to extract specific fields, such as .address from the example output, or .address and .label for two fields. Because the output from listreceivedbyaddress is a JSON array of multiple addresses, the necessary jq command is more complex:

```
$ dogecoin-cli listreceivedbyaddress | \
  jq '.[] | .address + " " + .label'
"DAY5wN... Consulting fees for Client A"
"DAyayA... Twitter tipbot"
"DDDKFs... Mining Pool"
"DFhv7M... GitHub address"
"DFpN6Q... GitLab address`"
"DJRU7M... imported"
```

Pay close attention to the jq section; this essentially says "At the root of the JSON provided (.) is a top-level array ([]). For each item in that array (|), get the .address and .label elements, then concatenate them together with a blank space between them (+ " " +)"[16].

While the *details* may be more complicated than throwing everything in a spreadsheet and pivoting to find all unique items in the Addresses column, once you get the hang of manipulating JSON output at the command line, it's fast to write these things.

What Can You Do With This?

Searching and filtering transaction data can be useful, especially if you need to perform accounting operations (see Generate Transaction Receipts, pp. 93). When did Client A pay you? Did they pay on time? What tax year? Ad hoc queries are easy here.

If you need to import this data into an accounting system and/or you don't want to balance your books on the same machine where your Core runs, export this data. CSV output works well with structured data systems which think of transactions as line items.

If you're performing data analysis or automating other tools or systems, JSON output gives you flexibility. In this specific case, the aggregating behavior of the RPC

---

[15] See https://stedolan.github.io/jq/ but, as always, verify with other places that this is still correct.

[16] Your author had to look up the appropriate syntax here. Search for jq examples online.

command may or may not work in your favor. Sometimes it's great, though, and there are always other wallet-manipulation and query commands available. Fortunately, you have multiple ways to get this information, depending on the tools you have, what you're comfortable doing, and what you're trying to accomplish.

Understand the Risks

Besides the normal risks of using RPC commands against any node, especially one with a wallet, be careful not to expose your addresses to anyone you don't want to have your addresses. That's why the addresses in the example figures here are obscured–not because your author doesn't trust *you*, but your author doesn't want you to trust *other people* blindly.

Keep your data safe and secure–all of it.

  # Tip #46   Generate a QR Code

If you take Dogecoin payments in some kind of in-person setting, you probably want an easy way to allow people to send you payments without talking through an entire wallet address out loud[17].

This is even more important when you want to remove some degree of human intervention from the process. Suppose you have a pinball arcade and you want to allow people to pay with quarters *or* with Dogecoin (see Manage a Dogecoin Arcade, pp. 301). How can you accept payments with as little friction as possible?

A Bitcoin Improvement Proposal has the answer.

Dogecoin-Aware Wallet Links

As defined in BIP-21[18], a hyperlink starting with `bitcoin:` and followed by an address tells your computer, phone, tablet, or other device to open some sort of wallet application and prepare to send a transaction to the given address.

Dogecoin adopted this obviously good idea.

To sell electronic copies of a book about Dogecoin online, you might create a BIP-21 address such as `dogecoin:DAY5wNkebzEyqUXCkN9koKNBuzXRKRTjcL` to receive funds. If someone clicks on that link on a device that understands that

---

[17]"That's D as in Doge, K as in Kabosu, no I don't want to say Donkey Kong, just trust me."

[18]See https://github.com/bitcoin/bips/blob/master/bip-0021.mediawiki.

format, their wallet will prompt them to decide whether to complete the transaction.

There's not much to BIP-21 you have to know besides "start with the prefix dogecoin:" and "follow it with the address you want to use to receive funds". If you're using another network, make sure your Litecoin or Pepecoin or Dingocoin or Bitcoin address is correct and that you're using the right prefix.

That works well on a web page, but if you're in an amazing Doge-aware pinball arcade, do you really want to pull up a webpage and search for the right link when you turn the corner and see the incredibly rare Baby Pac-Man! machine you weren't sure if you imagined in a dream?

Wouldn't it be more fun to snap a pic and then start playing?

Dogecoin-Aware Images

You're probably already familiar with QR codes, the three-dimensional barcode looking images that contain arbitrary data and take you to real estate listings or menus in restaurants that don't want to hand out grubby menus.

You may already have seen that the Dogecoin Core creates both a BIP-21 address and a QR code when you ask it to produce a receiving address. That's not always 100% convenient. What if you're using an offline wallet? What if you don't have the Core available? What if you don't have the Core GUI running?

Fear not; you can make your own QR code with your own address!

While multiple websites purport to generate QR codes for you, you shouldn't have to trust a third-party to do it right. While you can and should double-check with your own phone or device that the code contains the link you want, it's important to be able to use these tools on your own too.

Many programming languages can generate QR codes, including Python with the module qrcode[19]. If you already have Python 3 installed, use python3 -m pip install qrcode. Otherwise, download an installer from https://platform.activestate.com/chromatic/Python-QR-Codes for Linux, Mac, and Windows systems.

This module includes an easy to use command-line utility called qr:

```
$ qr 'dogecoin:DAY5wNkebzEyqUXCkN9koKNBuzXRKRTjcL'` \
> book-address-qr.png
```

This will write a PNG file as output containing the QR code encoding the address. If your terminal supports image output, you can test the command by skipping the

---

[19]See https://pypi.org/project/qrcode/ for more details.

file redirection and pointing your phone at the screen, as shown in Generate a QR Code from the Command Line, pp. 148.

Figure 6.5: Generate a QR Code from the Command Line

Understand the Risks

A QR code lets you take payments from multiple devices with reduced human interaction. That makes the link in the QR code controls the payments users make. Double-check that the embedded link is a BIP-21 link points to *your* address.

If you print these QR codes to attach to your pinball machines, be sure to do so in a way that doesn't hurt the paint. Attach them to glass or metal surfaces with the appropriate adhesive. You'll thank yourself later.

Printing QR codes runs the risk of address reuse. While it's safer to generate new addresses for each transaction, balance the risk against increased human interaction every pinball play. Consider frequent address rotation (see Rotate Machine Addresses, pp. 307) and generate new QR codes to match (weekly or monthly).

Finally, be aware that a payment system that relies on people scanning codes with their phones is susceptible to the risk that someone may print out their own QR codes and stick them over the top of your own codes, sort of like an ATM skimmer. Watch your machines carefully.

  Tip #47   Customize Your QR Code

Your pinball arcade is doing great (see Generate a QR Code, pp. 146), but there's something missing. Your branding feels generic. You want more sizzle, besides the flashing lights, the satisfying thunk of a steel ball on a waxed playfield, and that horrible knocking sound whenever someone wins a free credit.

How about customizing your QR codes? Add your logo or a meme or something otherwise exciting, surprising, or attractive to make your payment experience more magical!

Write a Little Custom Code

You need two Python libraries for this, qrcode and Pillow. If you installed qrcode yourself, run python3 -m pip install "qrcode[pil]" to install Pillow. Otherwise, check your package manager manages Python for you. With everything installed, write this little program:

```
#!/usr/bin/env python3

import qrcode
from qrcode.image.styledpil import StyledPilImage
from qrcode.image.styles.moduledrawers.pil \
    import RoundedModuleDrawer
from qrcode.image.styles.colormasks import RadialGradientColorMask
from qrcode.constants import ERROR_CORRECT_L
from sys import argv

def main(address, logo):
    qr = qrcode.QRCode(error_correction=ERROR_CORRECT_L)
    qr.add_data(address)

    qr.make_image(
        image_factory=StyledPilImage,
        module_drawer=RoundedModuleDrawer(),
        color_mask=RadialGradientColorMask(),
        embeded_image_path=logo,
    ).save("qr" + address + ".png")

if __name__ == "__main__":
    main(argv[1], argv[2])
```

Run this with arguments something like:

```
$ python3 custom-qr-code.py DAY5w... images/chromatic-shibe.png
```

Figure 6.6: An Eye-Popping, Bedazzling QR Code

This will generate the QR code shown in An Eye-Popping, Bedazzling QR Code, pp. 150, with the QR code for this book's homepage and the fantastic junior developer shibe puppet from Put Your Face on Your Wallet, pp. 109. Print, share, send to an e-ink display, paint on a wall, do whatever you want with this–but remember, if you use *this* logo, any funds go to the author's wallet!

Understand the Risks

This code enhances the `qrcode` image to round the image's edges, add a color gradient to the QR code, and embed your logo in middle of the image. The result gets written out with a predictable name, suitable for printing, including in a website, et cetera. The more complex you make your QR code–especially if you use a large or complex additional image–the greater the chance that a QR reader will fail to read it appropriately. Test your image on multiple devices for looks *and* behavior.

  # Tip #48   Work Without a Wallet

As other tips in this chapter remind you over and over, your wallet contains your private keys and they deserve your highest protection. Securing access to your wallet, securely backing up your wallet, and keeping people out of your wallet all contribute to the security of your private keys and, by extension, the security of your Dogecoin.

Sometimes the safest thing you can do with your wallet is *not* to use it.

Not Using Your Wallet

The Dogecoin Core has a configuration option called `disablewallet`. If you set this in your *dogecoin.conf* file or pass it on the command line, the Core will skip loading or generating a wallet and will disable certain features. This command works whether you run the GUI or the command-line server-only program *dogecoind*.

What Happens When You Disable Your Wallet

Remember that a wallet is just a way to access your private keys. When you use this option, all of your addresses will still receive transactions and you can still send transactions from them (if you do so from a system that has access to your wallet). You can still monitor transactions across the network. Your node will still connect to other nodes, transmit and receive blocks, and keep your local storage up to date.

Any notification features you have enabled, such as transaction actions (see act_-on_wallet_transactions>) will not execute because there's no wallet. You can use *most* RPC commands that work with the blockchain: examining blocks, deciphering transactions, working with the mempool. You cannot, however, use any RPC commands for working with wallets because there's no wallet enabled! This also means that no one who can get control of your node by any mechanism can transact on your behalf.

151

When to Disable Your Wallet

You might want to disable wallet features entirely in two or three scenarios.

First, if you're running a node on a computer you don't have complete access to at all times. You might run it on a computer in a friend's basement or in a virtual machine launched with a cloud provider somewhere. If the value of the node to you is that it's participating fully in the network, then you don't need a wallet enabled.

Second, if you're doing some development or testing work. For example, if you decide to hack on the core, play with a development release to give feedback, or explore a potential upgrade, working with un- or under-tested code represents a risk. Keeping your wallet separate is a good idea.

Third, if you want to be extra-special safe and secure, you might connect your wallet on a separate filesystem (a USB key, with proper backups of course) *only* when you want to *send* a transaction. If you usually run *without* a wallet enabled, the potential for wallet compromise due to user error or malicious behavior is lessened.

Understand the Risks

What's the difference between this and *not* using your real wallet with your real keys on a Core node? In effect, very little. In practice, if you enforce the habit of running with your wallet only when you care about working with your actual wallet, you'll have to do the extra work of enabling your wallet when you're ready to use it.

In other words, you will break yourself of the habit of assuming your wallet is always active and ready to go.

Of course, every node with your wallet attached (or even on disk somewhere) is a node that has access to your wallet and probably a computer where your wallet could be compromised. Even if you've encrypted it, it's still more secure if *no one* has access to it than if someone does.

Are there ways to get the best of both worlds? Yes, with their own tradeoffs! Still, if you look at a wallet as a vulnerability (because it is), you can measure the pros and cons of giving a node access to it and evaluate each situation as your needs change.

Tip #49   Export and Extract Wallet Data

A wallet is good for keeping your keys and addresses and labels together. In fact, you can't do much to send or receive Dogecoin without a wallet somewhere. A node without a wallet can do a lot for the network, but it doesn't do much for you personally.

A previous tip explained how to back up your wallet (Securely Back Up Your Wallet, pp. 136). This approach works great if you want to[20] keep multiple secure copies in multiple places just in case something happens to your main system. A wallet backup, if stored securely, can give you the ability to recover from a disaster like a goat chewing through your laptop's power cable and bricking your hard drive, but it gives you little more than redundancy. It treats your wallet as opaque, obscuring its *contents*: your private keys, HD key derivation paths, labels, and addresses.

A plain-text version of your wallet gives you more options–if you handle it with caution. If you need to operate on this set of data as a whole, exporting your wallet makes this possible–at the risk of exposing your wallet data in plain text.

Understand the Risks

This tip will produce an unencrypted, plain-text representation of your wallet. The contents of this file (and the data in your wallet in your nod) shouldn't be intimidating, but your secrets are most secure when they're kept secret. This file will contain your private keys *and* the Dogecoin addresses derived from those keys. Your private keys will be in a file on your computer in plain text, and anyone who can get their hands on that file will also get their hands on those keys–the same way that anyone who gets their hands on your unencrypted wallet or unencrypted wallet backup has your keys and can spend your Dogecoin. If you have a cool 100 million Dogecoin in unspent transactions received by an address, you–or anyone else–can look at this file, find that address, and see exactly which private key you need to use to *spend* those coins.

If you use the knowledge in this tip, use the data in your exported wallet carefully, then discard of the file effectively. Merely deleting it may not be enough; on Unix-like systems such as Linux, a BSD, or Mac OS X, use a binary called shred to mangle the file past the point of hard drive recovery before deleting it. On

---

[20] . . . and you do want to! .

Windows, look for a reputable file shredding utility. Even so, never underestimate the lengths to which someone might go to get your private data.

Dump Your Wallet

The Core RPC command `dumpwallet` writes the contents of your wallet in a specific format to a file specified as the argument to the command. Using the console (Command the Core, pp. 42) or `dogecoin-cli` or another RPC mechanism, send the command `dumpwallet insecure-wallet-shred-now.txt`, then navigate to the file. You can read it from there.

> ### Secure Backups with Dogecoin Core
>
> As of version 1.14.6, Dogecoin Core stores all backups and exports to a specific backups directory. Unless you've overridden the default location, you can find these files in the *backups* subdirectory of the *.dogecoin* configuration directory.
>
> For previous versions, you can give the full path to the location on disk where you want the Core to write your file.

Inside Your Wallet Dump

When you look at this file, you'll see it's long[21]. There's a short header giving information about the version of the core used to dump it, the timestamp of the dump, and the most recent block the Core had seen at the time of the backup.

Then you'll see the private masterkey for HD addresses (see Use a HD Wallet, pp. 39), and finally one line for each of your wallet addresses, something like (lines truncated for brevity):

```
QVcp9S... 1970-01-01T00:00:01Z label=2022%20Consult # addr=D7o...
QAdcVL... 1970-01-01T00:00:01Z label=2023%20Consult # addr=DBx...
QUt9eE... 2023-04-18T01:02:13Z label=            # addr=D8P...
QQgGGm... 2021-04-19T22:42:29Z reserve=1         # addr=DP7...
                               hdkeypath=m/0'/0'/7'
```

This file contains multiple fields, each separated by spaces, with an optional comment at the end of the line preceded by the # character. The first field is the private

---

[21] Your author's file is more than 450,000 lines long.

key. The second field is the timestamp when the private key was added to the wallet. In the third field, you may see a `label=` (though the value may be blank, as it is in the third line of this example). You might also alternately see `reserve=1`, which indicates that the Core has generated this address for future use, but nothing has used it yet.

Finally, the comment may include the public address associated with the key. When the Core has derived the key from an HD seed, you will also see that key's derivation path in `hdkeypath`.

Extracting Data From Wallet Dumps

After several lines of header information, each private key and address has one line in the file. If you can read this file, you can get every address out of it. This sounds like a job for automation, so warm up your favorite programming language!

The *easy* way to work with this data is to assume you already have a dump file somewhere on your system, then read it line by line and extract the data you want. In pseudocode:

- loop over every line in the file

- skip blank, header, and comment lines (headers and comments look identical)

- split all other lines on space characters into into data chunks

- extract address data from the appropriate data chunk

In the programming world, a *regular expression* (or *regex* for short) is a concise description of the shape of data, like "A capital letter followed by multiple numerals or letters, up until but not including a space character". They're very useful, but if you're not familiar with them, they can be intimidating. That's okay. You don't have to understand them to use this tip!

Regex-wielding programmers[22] can do this in a couple of lines. Look for a chunk of data starting with `addr=`, then capture everything after that until the next space. With Perl-style regular expressions, write something like `/addr=(\S+)/` and, if the match succeeds, take the address from the first (only) capturing group.

Once you have this address, you can write it to a file which you can use for other purposes.

---

[22] With no apologies for https://xkcd.com/208/.

Automated Extraction From Wallet Dumps

This approach has two drawbacks. First, you have to write the code yourself. Second, you have to manage dumping the wallet, processing the file, then remembering to remove the wallet dump securely.

Another approach is to automate the process: make the RPC call, loop through the plain-text wallet dump, extract the addresses, write them to a separate file, then delete the original file. This is a variant of the technique used to wrap RPC calls with other behavior (see Enhance RPC Calls, pp. 80).

Fortunately, your author has already written code to do all of this with version `1.20230424.0253` of `Finance-Dogecoin-Utils`. The `dogeutils` command has an option `exportaddresses`:

```
$ DOGEUTILS_USER=lisa dogeutils exportaddresses > my_address_list
```

As with the previous `dogeutils` uses, you need your authentication for RPC configured correctly. Also be sure to redirect the output of this command to a file, as you may get thousands or more lines of output. Behind the scenes, the code handles making the wallet export for you as well as deleting the exported file.

Also review the code before you run it. Even if you trust the author, you should verify that the code came from a trusted source, has no obvious or non-obvious bugs, and seems to do what it claims. Any code you download and run on your private information is a risk.

What Can You Do With This?

You can do at least three other things with this data:

- Skim it and see what you can learn from reading. There's a lot here! This can be a good springboard for doing more research.

- Forget you've exported it, give the laptop to your cousin who wants to study computer science, and wake up in the middle of the night realizing that everyone in University may have been able to grab this file off of an unprotected network share.

- Process this file for another purpose, then delete it securely.

To the last point, you can do a *lot* more with a wallet dump than export addresses. You could back up your private keys (especially if you write them to a file and encrypt it). You can find addresses without labels and figure out labels for them (and add labels to them). You can look for addresses waiting in reserve or addresses already used.

This file doesn't contain a large amount of unique data, but it contains data you can't easily access in other ways, so if you have any needs listed here, this is your best approach.

  # Tip #50   Watch Wallet Addresses

One of the complicated (and valuable) features about blockchains is that all transactions are public. Anyone, anywhere can review any transaction and trace the lineage of its inputs back to one or more coinbase transactions. They may have to put work into it, but the coherence of the entire chain is observable.

Security is important when it comes to your own computer too. While it's most secure to run a core node without a wallet attached (see Work Without a Wallet, pp. 151), sometimes you absolutely must have a Core node with knowledge of the transactions that are important to you–transactions where you care about sending or receiving Dogecoin.

You run two risks with a wallet attached to a Core node. First, anyone who can access your node has a better chance of being able to get your private keys, and thus the ability to spend your unspent transactions. Second, anyone who can see the addresses you care about may be able to track your transactions back to you.

Securing your computer overall helps with both risks. Working without a wallet addresses the first risk, but it makes the second risk moot by taking away your ability to track interesting transactions.

Fortunately, there's a way to have the latter while reducing the exposure of your private keys.

Watch-Only Addresses

Dogecoin Core borrowed a Bitcoin Core feature where you can import an address to your wallet. You can't send or receive funds from or to it, but the Core will index all transactions to and from the address and let you query them. A *watch-only address* is an address in your wallet–an address for which the Core tracks transactions–without an associated private key. You may have the public key or the address only, so the Core knows that you care about all transactions to or from that address. Without the public key, you have no way of spending any funds sent to that address.

This can be handy when you care more about tracking than you do spending. For example, if you run a pinball arcade or charge people 5 Dogecoin for admission into your goat petting farm, your ability to see transaction confirmation as soon as possible is more important than your ability to spend immediately. You can always use a full wallet later when it comes time to buy pinball wax or goat food.

What's the difference between watching an address and watching a key? Remember the order in which you can derive data! Given a secret key or passphrase, you *can* derive a private key. You *cannot* go backwards. Given a private key, you can derive a public key. You cannot go backwards. Given a public key, you can derive an address. You cannot go backwards.

Watch-a-Bunch

If you've dumped your wallet and extracted a bunch of addresses (see Export and Extract Wallet Data, pp. 153), the rest of the tip might seem easy. Before you go further, you must understand that the Core must process a transaction for your wallet to know that a transaction has affected it. How do you know which transactions to process? You have to examine every block containing a relevant transaction. How do you know which blocks are relevant?

Either you already know this, or you need to ask the Core to rescan enough of the chain to pick up every transaction you care about.

The `importpubkey` RPC command adds a public key to a Core wallet in watch-only mode. Any addresses derived from these public keys get watched for transactions, but you can't spend any unspent transaction inputs until you create new transactions with the appropriate private keys.

This command has one required and three optional arguments. You *must* provide a valid public key. You *may* provide, in order, a string label to attach to the address of this key, a true or false value whether to rescan the blockchain after importing the key (defaulting to true, always rescan), and (as of Dogecoin 1.14.7), an integer value representing the height of the block from which to start scanning. If you provide no height but do request a rescan, or if you're using an earlier version of the Core, this command will scan the entire blockchain from the genesis block with height 1.

This can take a while.

If you don't have a public key but you do have an address–such as the Dogecoin Developer tipjar[23], as shown at https://foundation.dogecoin.com/announcements/ 2022-12-31-corefund/, you can watch that with the `importaddress` command, which behaves similarly to the public key import command.

---

[23] 9xEP9voiNLw7Y7DS87M8QRqKM43r6r5KM5

```
$ dogecoin-cli importaddress 9xEP9voi...
```

When the rescanning process finishes, if you asked for it, you will see all transactions to the watched address as if they were transactions to your own wallet
addresses. Any RPC command or other query you make of the Core can see these
addresses and transactions. Look for parameters to these queries to include or exclude "watchonly transactions" or "watchonly addresses" to disambiguate funds
under your control from funds associated with addresses you imported.

What's the Right Height?

To be extra safe, you can rescan from the genesis block which has a height
of one. That's guaranteed to give you good answers. To do less work, use
a block explorer to see the *earliest* block in which a transaction to a given
wallet occurred.

If you've extracted your own wallet addresses, use the block height of
your oldest transaction. You'll save yourself computing power and time:
scanning the older blocks will find nothing relevant to your interests.

Script Your Watching

Given a list of wallet public keys, you can script a series of commands to import
every address. There are a couple of approaches:

- Write a program to loop through the keys and make RPC calls directly
  against a Core node

- Write a program to loop through the keys and invoke 'dogecoin-cli' for each
  key

- Write a program to write a list of commands and copy and paste that into
  the Qt Debug Console (see Command the Core, pp. 42)

Any of these approaches will work. All are straightforward depending on your
preference. You can do the same thing with wallet *addresses* too, if you swap the
RPC command for importaddress.

Understand the Risks

Be mindful of a couple of things. First, any address or key you add to a watch-only wallet reveals something to anyone who can get access to your node. Sure, you and a few hundred thousand of your closest Doge-friendly friends may be watching the developer tipjar or suspected whale wallets to see what's going on, but if you've set up a consulting operation and you don't particularly want the general public to see that your hard work has netted you a sweet 10k Doge windfall, consider what kind of security you want to have on hand to avoid leaking information.

Even though you can't *spend* these funds (because the wallet doesn't have the private keys), anyone who gets your wallet can see what you're watching. Keep your files safe and private!

Also remember that rescanning the entire blockchain can take time and resources, and has a good chance of preventing you from doing other work with your node until the rescan finishes. Take advantage of the optional rescan height argument, when possible.

Furthermore, every address or key you watch will take up space in your wallet. If your wallet size gets too big to back up encrypted on a USB key (see Securely Back Up Your Wallet, pp. 136) because you watched every address you could find, you'll have to find an alternate approach.

Alternately, skip rescanning if you're importing a lot of addresses or keys by explicitly providing a false parameter to the rescan option. When you've finished importing everything, use the `rescan` command explicitly with the lowest block height that will cover everything you've imported. This command is new to Dogecoin Core as of 1.14.6, so you can feel a little special knowing that you have a feature other similar blockchains either don't have or borrowed[24].

  Tip #51   Replenish Your Address Pool

If you're regularly receiving transactions–perhaps doing a lot of consulting, selling things online, or otherwise doing something interesting, you'll need a lot of addresses. While other tips demonstrate ways to automate this process, that work can be more suitable for running some kind of e-commerce system than if you're comfortable grabbing an unused address from the Core a couple of times a week or a couple of times a day.

---

[24]They're welcome to borrow it! Isn't it nice to know that Dogecoin did something to make everyone's lives easier?

Fortunately, Dogecoin Core maintains a pool of unused addresses so that you always have fresh addresses available. By default, this pool has 100 addresses in it. When you use the last one, the Core will generate more.

This is under your control. While you don't have to do anything to keep a steady supply of unused addresses available, you *can*–especially if you're going to be going through a lot *or* if you make a lot of backups and want to be sure your backups contain all of your addresses.

Understand the Risks

If you're not running an HD wallet (see Use a HD Wallet, pp. 39), keeping a key pool around is *essential*. Deterministic address generation from an HD wallet removes much of the risk that you'll lose addresses if you generate new addresses and miss them between backups.

With all that said, *should* you use this feature? There's little risk from doing so, as long as you keep your wallet backups safe and secure (see Securely Back Up Your Wallet, pp. 136). Consider this one more reason to upgrade your wallet to take advantage of HD features. The Core's key management approach before deterministic address generation had risks. Keys were generated by random numbers and addresses produced from those keys. You can see how the new key management and derivation system mitigates those risks.

Key Pool Refill

The keypoolrefill RPC command tells the Core to generate new keys. With no argument, it'll generate 100 more. If you provide a value, it'll generate that many more.

Use the getinfo RPC command to see the current size of the keypool; look at the keypoolsize entry in the returned object.

One final note: the keypool configuration option set in your *dogecoin.conf* file or specified on the command-line changes the default number of entries in the keypool. If you choose a value other than 100, use keypoolrefill to ask the Core to fill the pool up to that number (assuming you choose a higher number).

## CHAPTER 7

Basic (and Advanced) Transaction Stuff

Hoarding coins can be fun for a while, but eventually they have to move around. That's what makes coins into a *currency*, after all: you trade them for goods or services. Whether that's buying a cute stuffed animal or flag online, getting paid for consulting or writing something funny/sad/insightful, or giving a tip to someone who made you smile, transactions are the beating heart of Dogecoin.

This chapter walks through smart, surprising, and silly things you can do with transactions.

Tip #52 Bundle and Track Transactions

While cryptocurrency has only been around for a few years, cryptography has been around for longer. People generally remember "Oh yeah, that Roman salad guy invented a cipher[1]!", but they may not realize that the other important foundation of cryptocurrency comes from accounting.

Accounting?

If you think of the Dogecoin blockchain as a transaction ledger[2], then every transaction should satisfy the fundamental accounting equation[3], where a credit on one side exactly equals a debit on another side.

Explained another way, if you send me 10 Dogecoin, my wallet gets credited by 10 Dogecoin and yours gets debited 10 Dogecoin. You have to have 10 Dogecoin available in your wallet somewhere, otherwise the transaction is invalid. Those 10 Dogecoin you have available had to come from somewhere else, or the transaction

---

[1] See https://en.wikipedia.org/wiki/Caesar_cipher

[2] The name gives it away already, doesn't it?

[3] See https://en.wikipedia.org/wiki/Accounting_equation

is invalid. You can trace them as far back as the point at which they entered the Dogecoin network altogether, and the numbers must pencil out at every transaction, otherwise your attempt to tip me is invalid and the network will reject that.

There's a lot of detail there, and a lot of other tips in this book explore part of the details and several of the implementations. You don't have to understand this fully to use Dogecoin, but to go deeply into how transactions work, you need to know the underlying principles.

Every time anyone or anything validates a transaction, they have to answer two questions. First, do these coins exist in the network? Second, have they already been spent?

Thinking in Transactions

What does it mean that you have 10 Dogecoin available to tip me or anyone else? You must have:

- an address in your wallet, which

- has received Dogecoin from other addresses, that

- add up to at least 10 Dogecoin, which

- haven't been sent to other addresses outside of your wallet, which

- would appear in confirmed transactions

If you think about this as gold coins in your backpack in a video game, you're not going to buy that cool horse with racing stripes if the vendor wants 100 coins and you have 0, or 99, or any number in between. Unless you can hand over 100 gold coins exactly, you're going to walk to the next castle instead of riding in style. The gold coins in your backpack are unspent coins. As soon as you hand them over, they're spent and you can no longer spend them.

Think about how you acquired all of those gold coins. Maybe your character received an inheritance from a distant great-uncle. That's 23 gold coins right there. Then your character spent a week in-game mucking the Augean stables for another 10 gold coins. You convinced a dragon to leave a town alone and migrate to a deserted island and received 7 gold coins as a reward, etc and so on.

Every time you earned a pile of coins from a quest, that's a transaction. Someone opened their pouch, counted out some coin, and handed it to you and you put it in your backpack.

Unspent Transactions

The blockchain ledger works kind of like your backpack full of gold, except every distinct batch Dogecoin you have received has a little note on it that says where it came from. In the video game, this would be like the owner of the stables handing you a little pouch of 10 gold coins with "For mucking these dirty stables" stamped on it, and then you put the little pouch in your backpack.

What happens if you stop at an inn and really want to sample the local pumpkin cider, but it costs one gold coin per pint? You open your backpack, rummage around for a pouch with at least one gold coin in it, and you break it open. Sorry, pouch labeled "Picked seven carrots from the farmer's garden in the starting village", you have a powerful thirst.

In this analogy, your backpack full of small, labeled pouches are unspent transactions. That's what you have to work with if you want to *spend* your Dogecoin. You don't have a backpack full of 100 gold coins; you have a backpack full of smaller pouches, and the count of all coins in those pouches adds up to 100 coins.

Every Coin Goes Somewhere

It's obvious how paying for your tasty beverage works if you can find a pouch with one gold coin in it. What happens if the smallest pouch you have has two coins?

You have a few choices. One, go thirsty. Two, hand over the pouch with two coins and drink two pints. Three, break open the pouch and make *two* new pouches. One of them goes to the friendly brewer and the other goes back in your backpack. The new pouches each get a label that reflects that you're paying for pumpkin cider.

The same would happen if the smallest pouch you had contained 100 coins or 1000 or however many. You'd make one new pouch for the merchant and one new pouch for you. In every case, the label for the new pouches would also includes the fact that these two pouches came from the previous pouch which you created for a reason, such as payment for picking carrots for a farmer (see Forge a Chain, pp. 7).

What's Really Going On

It sounds like a lot of bookkeeping to keep a backpack full of pouches of coins, so the analogy breaks down there. If you really carried a backpack full of gold, besides the obvious weight and robbery risks, you probably wouldn't bundle them in pouches, because you don't really care where they came from (they're in your backpack now) and you can't really double spend them (once you hand the brewer your gold coin, you're getting your cider).

Because we're *not* dealing with physical items here, just numbers in computers, we do need additional ways to control our transactions to provide those two guarantees. Hence the ledger concept, the accounting controls, and the labels on all transactions.

All of these mechanics give rise to other implications the rest of this book explores directly or implicitly. There's one nuance you should be aware of, however: it costs time and energy to verify that your transactions are what you claim they are and that they're valid. Every Dogecoin transaction you make has a tiny fee attached to it to pay the people who verify that you can do what you're trying to do. This is very important and will come up again soon.

  # Tip #53   Inspect a Transaction

Holding Dogecoin is interesting and spending Dogecoin is fun, but a whole world of things happens when you do either. The fundamental concepts of blockchains require that everything happens in public, in a verifiable way: mining coins, sending coins, receiving coins, et cetera. The *mechanism* of your actions is visible to you and everyone.

Understanding what happens when you send or receive coins can open the door to a lot of possibilities.

What's in a Transaction?

To understand a transaction, think of a ledger. All the bookkeepers and accountants reading this book can now say "Yes! Finally!" and enjoy the warm glow for a second. In the simplest terms (and all those bookkeepers and accountants can say "Oh no, it's not *that* simple!"), a ledger tracks a few things:

- Money entered the system at a specific point in time

- Money changed hands at a specific point in time

- Money exited the system at a specific point in time

That repeated phrase "at a specific point in time" is essential. At every part of the process, after every change to the ledger, the numbers should all add up correctly. If your author takes a transaction worth 1000 Doge and sends you 900 and pay 0.01 in transaction fees, the other 99.99 must go somewhere else, otherwise the transaction is invalid and the ledger is incomplete.

Thus, transactions have inputs and outputs–and other data.

Transaction Inputs

Transaction money must come from *somewhere*. For you to send the author 100 Doge to buy a copy of this book, you must have access to at least 100 Doge. With all of the addresses under your control in your wallet, you must find enough unspent Doge to add up to that 100 Doge threshold, plus enough to cover the transaction fees.

Make note of the word "unspent" there. If, per the accounting guidelines earlier, every ledger in the entry balances perfectly, we should be able to examine every address under your control and see the exact inputs and outputs and get a perfect balance of the funds you control but have not already spent. That's the point! On the blockchain, miners verify that the transaction you're attempting to transact is coherent with regard to the blockchain itself.

To make life easier for miners, a valid Dogecoin transaction requires the sender to identify the source transaction of every coins used in the transaction. It's like when you swipe or tap a payment card at your favorite falafel stand; it's not enough to say "I'll send you the money". You have to identify the source of the funds.

Clever readers may be asking "What prevents someone from picking an arbitrary transaction as an input and spending someone else's koinu?" Transaction inputs also have special data that proves that whoever created the current transaction can satisfy the conditions to unlock the input transaction(s). There's a lot to discuss here, so other tips will cover this in more detail.

The Ultimate Input

Where do miner rewards come from, if transaction fees are so small? Where did the initial Dogecoin come from? The first transaction of every block is called a coinbase transaction. This special transaction type can create more Dogecoin out of thin air. Every non-coinbase transaction must trace its inputs back to one or more coinbase transactions.

When you're looking at a transaction[4], you'll see one or more inputs, except for coinbase transactions. This makes ledger bookkeeping work correctly.

---

[4] It's not a fireworks factory; we'll get there soon!

Transaction Output

Inputs and outputs go together just like taking a kruggerand out of your pocket puts it in your hand. Without an output, your transaction sits around, doing nothing.

A transaction can have one or more outputs. Each output contains its own data. First, it needs the number of koinu included. If you're paying 100 Dogecoin for a copy of this book, the output needs to contain 100 Dogecoin. Second, the output needs a condition under which the recipient can spend the provided Dogecoin. If you're sending this to one of the author's addresses, you need to provide a cryptographic puzzle that only the owner of the private key behind that address can solve.

That's not the only kind of condition you can produce, of course, but that's a topic for another tip (see Timelock a Vault, pp. 200, for example).

> ### Transaction Metadata
>
> Transactions also include bookkeeping information, such as the transaction ID. For now you can ignore most of that information, but you absolutely need the transaction ID to do anything interesting; it links every transaction in the blockchain together into a chain.

What's in a Transaction?

How about looking at a real transaction? If you have the Core running on your computer and if you have a wallet that's received inputs, use RPC commands to examine one of your transactions.

```
$ TX=$(dogecoin-cli listreceivedbyaddress | \
  jq -r '.[0].txids | .[0]' )
$ RAWTX=$( dogecoin-cli getrawtransaction $TX )
$ dogecoin-cli decoderawtransaction $RAWTX

  {
    "txid": "73dc8...",
    "hash": "73dc81...",
    "size": 225,
    "vsize": 225,
    "version": 1,
    "locktime": 0,
    "vin": [
      {
```

```
        "txid": "f806b..",
        "vout": 1,
        "scriptSig": {
          "asm": "...",
          "hex": "..."
        },
        "sequence": 4294967295
    }
  ],
  "vout": [
    {
        "value": 6.90000000,
        "n": 0,
        "scriptPubKey": {
          "asm": "OP_DUP OP_HASH160 3b31b1... OP_EQUALVERIFY ...",
          "hex": "76a914...",
          "reqSigs": 1,
          "type": "pubkeyhash",
          "addresses": [
            "DAY5w..."
          ]
        }
    },
    {
        "value": 978.55966000,
        "n": 1,
        "scriptPubKey": {
          "asm": "OP_DUP OP_HASH160 f1e517... OP_EQUALVERIFY ...",
          "hex": "76a914...",
          "reqSigs": 1,
          "type": "pubkeyhash",
          "addresses": [
            "DTC7q9..."
          ]
        }
    }
  ]
}
```

This example truncates some information to reveal the structure of a transaction without burdening it with details. This transaction has a single input (in vin), with a reference to a single input transaction (txid). This transaction also has two outputs (both found in vout). These outputs have a value of 6.9 and 978.55966 Dogecoin respectively. If you add these amounts together, you'll come up with almost the same amount as was received in the input transaction, less a few koinu used for transaction fees.

The command-line commands feed into each other (hence the use of shell variables). The jq line is only for this example; it looks at all of the addresses which have received Dogecoin inputs, looks at the first one, looks at the first transaction, and gets its transaction ID. You could use *any* valid transaction ID, including the transaction ID of the input transaction and its input transaction(s) and so forth, until you reach a coinbase transaction.

This is all easier using a blockchain explorer, but if you do it yourself you can trust and verify your own copy of the blockchain and not rely on someone else's opinion.

Understand the Risks

There's no risk in exploring the blockchain other than the same risk as always of running a Core node on a computer hooked up to a network. Issuing these RPC commands over the network (not on your local machine) has the same risks as always.

The biggest risk here is making assumptions about the structure of transactions and how they work. The "Mastering Bitcoin" book[5], by Andreas M. Antonopoulos, explains transactions in much greater detail. It's a good reference for anyone interested in the structure and use of transactions and Bitcoin-style technology in general.

  # Tip #54   Decode Transactions

If you find yourself doing something complicated more than once, consider how to avoid repeating yourself–especially if the complicated steps are easy to get wrong. This could be transcribing one set of data between systems or copying and pasting information from multiple processes, windows, or machines.

If you read the previous tip (Inspect a Transaction, pp. 166) closely, you noticed that the Core currently provides no direct way to go from a transaction hash to the decoded transaction, or a list of inputs or outputs, or output scripts, or anything else.

You can do the two- or three-step shuffle to get this data, but why do that more than once?

Chaining RPC Commands

Another tip discusses RPC command stacking (Enhance RPC Calls, pp. 80), where you feed the output of one RPC call into another's input. This allows you to act

---

[5] See https://amzn.to/3xxDKHf or https://github.com/bitcoinbook/bitcoinbook.

as if the Core supported only one call to perform a bunch of behavior, such as decoding a transaction into a data structure given only the transaction's hash.

If you were to do this manually, you'd have to call `getrawtransaction` with the hash, then feed the result of that into a call to `decoderawtransaction`. In the ideal case, this works perfectly.

If either RPC call fails, you need to account for that failure. That code will look something like this *pseudocode*[6]:

```
def decode_tx_from_hash( tx_hash ):
    tx := RPCcall 'getrawtransaction', tx_hash

    if tx.error != nil:
      throw tx.error

    decoded_tx = RPCcall 'decoderawtransaction', tx.result
    if decoded_tx.error != nil:
      throw tx.error

    return decoded_tx.result
```

You can write this in any language you like. Alternately, install the `dogeutils` library and toolkit[7], which does this for you.

Understand the Risks

The risks here are minimal. Adding extra calls through another language will add latency to the process, though you can add whatever error checking or convenience features you like if you provide your own implementation. This can be especially useful if you wrap calls which take optional parameters.

The greater risk is long-term maintenance. If the RPC calls you use change between releases, your wrapper will have to adapt to them. If the Core adds a call with the same name as yours, your code will continue to work, but you may see divergence between your code and the Core's implementation. Consider the maintenance costs of doing something different from the Core.

Relying on anyone else's wrapper, especially if it performs authentication, means giving up some control to someone else's code. Audit it yourself (or have someone you trust audit it very carefully for you) before using it.

---

[6]Did... did you just invent a weird new notation for this? Yep!

[7]See https://metacpan.org/dist/Finance-Dogecoin-Utils/view/bin/dogeutils.

Tip #55   Identify Input Transactions

Suppose you head to the local taco truck for lunch one sunny weekend. You buy three delicious street tacos and a cup of limonada. Your total is $17.95, including tip. You have $20 in your wallet. What happens next?

You reach into your wallet and pull out a $10 bill and two $5s, then give them that to the cashier. They hand you back $2. You wave off the nickel, saying "give it to someone who needs it". You put the $2 in your wallet and take your limonada to sit and wait for your tacos.

What just happened? Those three bills entered your wallet via one or more input transactions. You consolidated them to make a new, outgoing transaction. Because the exact values you had from the inputs didn't match up with the exact value you needed for the outputs, you put $2 in change back into your wallet–and you put a nickel aside for someone else.

This happens all the time in blockchain transactions. The amount of coins in all inputs used in a transaction must be completely consumed, whether the amounts are sent to recipients, returned as change, turned into transaction fees, or donated to miners.

If you're watching the addresses in your wallet because you're selling something (see Act on Wallet Transactions, pp. 194), you must distinguish between "someone put new money in my wallet" and "I took money out of my wallet and added change".

Suppose the term "input transaction" means "a transaction to one of your addresses from someone else", the term "change transaction" means "a transaction from one of your addresses which puts change into another of your addresses", and the term "output transaction" means "a transaction from one of your addresses to an address belonging to another person".

Restrict Input Transactions to Addresses

Sometimes the best solution to a problem is to avoid the problem in the first place. What's the problem here? If you're watching your wallet for new transactions that affect any of your addresses, you need to be able to distinguish between someone sending you new Dogecoin you didn't have before and your own wallet sending Dogecoin to itself.

If you have a wallet address that's never received a transaction before, any input to that address is a new transaction *to* that address; you can't send change from an address if it has no unspent coins. If you never, ever reuse addresses, you might be able to get away with the assumption that any new transaction to an address is a new input transaction.

The second most easy case is when you're careful to send change to a new address whenever you send a transaction *from* an input address. In other words, if you never send change back to the same address, you'll never see an input transaction to that address that isn't a new transaction. Whether you automatically generate wallet addresses or manually curate a list of available addresses (see Replenish Your Address Pool, pp. 160), segregating new addresses between "other people can send funds here" and "only I can send funds here" allows you to watch the former for inputs and ignore the latter.

You might not be able to make either of these assumptions safely, however. Fortunately, you have what you need to distinguish between input transactions and change transactions.

Look Inside Transactions

Given a transaction id, you want to look for a couple of characteristics (see Inspect a Transaction, pp. 166). First, your wallet address should be in the transaction's output list. Second, the transaction should have at least one input address that isn't one of yours. Ideally, none of the transaction input addresses will be your address in the output list.

To make this work, you have to decode at least two transactions. One is the input transaction which triggered your payment system. Given its transaction id (and knowing your address), look at the transaction's output list (the vout section of the JSON emitted from decoderawtransaction) to find your own address.

Then for every input transaction (in the vin section of the emitted JSON), look up and decode *those* transactions and look for your wallet address in their vout lists. The vout entry of the JSON for each input transaction will tell you the position of the previous transaction which added funds to the input address. In other words, given a decoded transaction like:

```
{
  "txid": "73dc8...",
  "hash": "73dc81...",
  "size": 225,
  "vsize": 225,
  "version": 1,
  "locktime": 0,
  "vin": [
    {
      "txid": "f806b..",
      "vout": 1,
      "scriptSig": {
```

```
        "asm": "...",
        "hex": "..."
      },
      "sequence": 4294967295
    }
  ],
  ...
```

...there's only one input transaction, with the transaction id starting f806b. Look in the vout data for that transaction, at position 1 (n is 1 in the vout array) and you should see the address as *receiving* funds.

This is a little bit complicated, so play with existing transactions using the RPC commands and/or a block explorer until everything makes sense. For now, remember two important things: reusing addresses makes your life more difficult and all outputs must be tied to inputs.

Understand the Risks

The biggest risk of misidentifying a change transaction or an output transaction for an input transaction is when you have an automated system for performing some kind of action on receiving a transaction. For example, if you sell a digital product–an ebook, access to a website, a subscription, images of cats wearing hats–you might not want to require human intervention every time someone wants your product or service.

While it might not be an *error* to unlock a subscription or send a download link multiple times, it's probably a mistake at best.

Configuring or building your payment system to be robust with regard to all of the types of transactions which may affect your wallet may take a little more work and understanding at the start, but it has the potential to make your life a lot easier and much less confusing as your needs and popularity increase.

  Tip #56   Host a Treasure Hunt

When you take the nature of the blockchain to heart, you soon realize that an address is just a number and a private key is just one way to derive that number. Similarly, a passphrase or mnemonic is one way to generate that private key.

In Dogecoin terms, an address in your wallet is a large number associated with a public key, which is itself a large number associated with a private key. Anyone can send to your address if they have it, but only someone with your private key can send any unspent koinu *from* your address. That's why keeping your wallet and private keys private is so important; anyone else with access to a private key has access to your funds.

A wallet's access isn't governed by access to a file on your hard drive. It's completely under the control of anyone who has the large number that's your private key, *or* anything they can use to derive that private key.

To prevent unauthorized access to that wallet address, keep your private key safe. If you *do* want someone to have access, you have to share it with them somehow.

That somehow can be fun.

A Big Whoop Treasure Hunt

Suppose you're a member of the local Chamber of Commerce on a tropical island in the middle of historical pirate country[8], and you want to encourage more pirate-related tourism. What could be better than a pirate-themed treasure hunt?

You probably don't want a bunch of tourists tromping around your jungle and back alleys and beaches with shovels[9], so you might want to stick with something a little controlled and a lot less muddy. How about a scavenger hunt? BIP-39 has the answer.

Generate A Mnemonic Phrase

The Bitcoin Improvement Proposal 39[10] proposes a way to take a series of words, ideally easy to remember, and use them to produce a private key which can produce a public key which can represent an address. As long as you have a deterministic way to generate these keys from the words, you don't have to remember the keys, whether writing them down, tattooing them on your eyelids, or saving them in an encrypted file on your backup hard drive. You just have to keep the words...

...or figure out a way to *remember* the words.

This is perfect for a scavenger hunt. Generate a phrase, cut it up into a riddle format, figure out the resulting public address, throw away the private key, send some Dogecoin to that address, and then wait for all of the amazing tourists to show up and try to figure out your riddle so they can solve the mystery of Big Whoop!

---

[8] Your author can already tell this example is going to get away from him, but the idea was too good not to try.

[9] Although this can be a good way to get the neighbor kids to help re-sod your lawn.

[10] See https://github.com/bitcoin/bips/blob/master/bip-0039.mediawiki.

If you read the BIP carefully between the lines you'll see the recommendation *not* to pick several words at random. There are carefully-chosen and curated wordlists available to reduce confusion and make it more likely that you *won't* lose access to your address.

How do you generate your phrase?

A hacker named Ian Coleman wrote some software called "Mnemonic Code Converter"[11] to generate phrases as well as derive keypairs and addresses. This is a single-page application you can run locally to manage this process for multiple spoken languages.

To use Ian's tool, download the page to your own machine, then open it in a web browser. Now comes the most important part: turn off your network. Turn off wifi. Unplug your Ethernet cable. Flip the power switch on your router. Put some salt on the coax or fiber or copper cable heading out to the street and let some goats chew on it. Whatever it takes to go offline, make sure that this web page can't give away what you're about to generate.

Is this too cautious? That depends! Keeping secret things secret is important, so think about what could go wrong, who you're trusting with what, and how to protect yourself even if you think you're safe. Multiple layers of safety are important.

Don't Stop the Hunt Before It Starts

Sure, you're hoping *someone* will figure out the passphrase eventually, but there's no fun if you accidentally use the wrong generator and some malicious operator gets the coins before you even publish your treasure hunting pamphlets. As always, check that the source you use to generate the passphrase is trustworthy. Ian's page links to alternatives, so if one gives you the howling fantods, switch to another–or compare the outputs from two against each other.

Generate a Keypair and Address

Now that you can generate a passphrase, you need to know a little bit more about how to turn this passphrase into a private key, a public key, and an address. While Ian's page can do all the calculations, it's still important to understand the details.

---

[11] See https://iancoleman.io/bip39/ or https://github.com/iancoleman/bip39/releases/latest/.

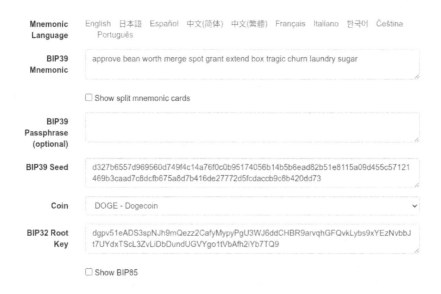

Figure 7.1: Generating BIP-39 Mnemonic and Dogecoin Keys

The smartest way to generate this key is to use hierarchical deterministic keys, as described in BIP-44[12] . This gives you access to a *range* of keys and addresses from a single starting point. We won't use that for this tip, but it wil come up again (see Use a HD Wallet, pp. 39).

Ian's tool uses a hidden calculation (read the source code) to turn that passphrase into a seed value which, when combined with the right parameters for the Dogecoin network, produces enough information to derive multiple keypairs and addresses. The HD chain for Dogecoin is `m/44'/3'/0'/0/0`. In other words, pick the *first* address from the Derived Addresses section of the tool.

Save the address (in this case `DEypUe2QRe1W4CsNsTQHpfiwMFLLM1zceQ` and the passphrase (approve bean worth merge spot grant extend box tragic churn laundry sugar).

Fund the Address

Now validate the address. On a *different* computer (or web browser) also not connected to the Internet, enter the passphrase exactly as written to verify that the `m/44'/3'/0'/0/0` address is exactly as you expected. If it is, great! If not, then

---

[12] See https://github.com/bitcoin/bips/blob/master/bip-0044.mediawiki.

177

you have a problem: someone who finally figures out all of your riddles all the way up to "sugar" can't easily go from these twelve words all the way to a private key they can import into their own wallet to gain control of the treasure.

This is important: some clue or rule will have to point people to how to *redeem* these coins. Maybe they bring the passphrase back to you, or maybe you point them at a QR code with redemption instructions.

When you're convinced that the keys and address are deterministically derivable from the phrase, send a thousand Dogecoin (or whatever the chamber of commerce can afford; don't skimp, because you brought in lots of tourist dollars!) to the address. Now the pressure's on.

Make a Fun Riddle

What can you do with these weird words? The order matters, so you'll have to figure out twelve riddles or places to hide these words. Maybe the first one is the first letter of each paragraph in the Welcome To Our Island pamphlet.

Avast!

Prove your worth.

Perhaps you will be the one to find our treasure.

Read carefully the instructions.

One by one you will uncover twelve secrets in all.

Very carefully the master treasure will reveal itself.

Enjoy our island!

Maybe the second one will have you trade the family's cow for a single seed used to grow a stalk to steal a giant's treasure. Possibilities exist.

Some phrases will be more difficult to manage than others; feel free to reload or regenerate the phrase until you feel confident you can produce a good scavenger hunt[13].

---

[13] "Worth"? That'll take some thinking. "Extend"? Awkward but doable.

Understand the Risks

This tip has discussed risks of using an external tool to help you generate keys and addresses. While the approach described could be more cautious than you like, the risks of malware are real. Take them seriously.

On an entirely different note, if you're generating a private key and address and throwing away the key, you should make sure of two things. First, you have to be able to recreate the address from the generated passphrase. Otherwise, those Dogecoin your chamber of commerce donated are gone forever-ish. Second, make sure that the generated address is actually a Dogecoin address. Use a Dogecoin address generator or verifier.

For a treasure hunt, if you expect people to participate without necessarily being fluent in Dogecoin or cryptocurrency, it might be wise to have a laptop or kiosk or tablet available in person where tourists can try to enter their passphrases to see if they can get the wallet address right. You *can* keep the address around for verification as long as you don't keep the passphrase or private key around; there's no harm in publishing that address. It's only derivable from the private key which is derivable from a passphrase.

Just remind folks not to step on the flowers and to leave their shovels at home.

  Tip #57   Vote on Goats

You've done it–fulfilled your life-long dream of buying land in the country where you can use your laptop while sitting on your back deck, sipping lemonade and admiring the view. Then wild blackberry vines grow and block your view. You can either put down your laptop and get out the gardening tools or adopt some goats and get back to sipping your drink[14].

The obvious answer is to adopt the goats. Four of them will do it. Now comes the difficult part: naming them.

Make a poll!

Blockchains and Voting

Blockchain technology has an interesting inherent property that transactions cannot be reversed. Once enough of the network has confirmed a transaction, it becomes increasingly expensive and troublesome to remove that transaction. In the case of Dogecoin, with blocks mined every 60 seconds or so, you can be pretty certain that a transaction confirmed ten minutes ago is permanent.

This property leads some people to suggest the use of blockchain technology for voting systems. While administering voting on the level of a town, parish, state, province, nation, or corporation is an interesting thought experiment, handling all the edge cases for something that matters more than capra hircus nomenclature gets complicated and messy.

Yet the allure persists. If you're clever and careful, you can use blockchain properties to validate properties of transactions to discern between valid/allowed votes and invalid/disallowed votes, depending on how you define all these terms.

Vote on Goats as a Poll

You've lined up a deal with a local herd to adopt four goats, on the condition that you take care of them appropriately for their natural lives *and* you immediately name them. You can't decide on one set of names, because you have several possibilities:

- Inky, Pinky, Blinky, and Clyde

- Mario, Luigi, Pauline, and Peach

- Sonic, Tails, Knuckles, and Amy

- Larry, Curly, Moe, and Shemp

- Link, Zelda, Epona, and Sheik[15]

---

[14]Recipe for blackberry lemonade: pick and wash blackberries. Add to lemonade. Muddle if you like.

[15]No spoilers; before you send angry feedback, remember they're hypothetical goats.

Why not use ninja turtle names? Too many syllables to yell when they wander away from the blackberry vines and into the roses.

For each set of possibilities, generate or choose a fresh, unused Dogecoin address. Then publicize your vote. To vote for the first set, send *your amount here* of Koinu to *selected address*. Go through the list. Tell people when voting starts and voting ends. Wait until voting ends, then pick the winner.

Counting Poll Responses

If you've done your job well, counting responses should be as easy as using the `getreceivedbyaddress` RPC call and adding up either the number of koinu sent to each address. The highest value wins!

What if, however, you start to wonder if allowing the wealthiest and most generous voter to have the most votes is ethical? It might not be! What if you want to count the *number* of transactions to each address and use that to decide on the winning names?

Use the `listtransactions` RPC call to list the transactions in your wallet, then use a spreadsheet or program or a friend with pen and paper to tabulate the results.

> ### Good Coders Borrow; Great Coders Steal
>
> Bitcoin's `listreceivedbyaddress` call has an additional fourth argument used to filter transactions to specific addresses.
> As of this writing, Dogecoin 1.14.9 does not have that feature. Watch for it in future releases!

With data on *all* of the transactions your wallet has received, you can also check the time of the transaction (the `blocktime` element of the data produced by `listreceivedbyaddress`) and verify that the transaction occurred within the voting period you chose. Anything early or late may or may not count, depending on your preferences.

Remember that Dogecoin records time in epoch time: the number of seconds since the birthdate of Unix on January 1, 1970. Use an "epoch time converter" built into your favorite programming environment or one of multiple web sites to translate these numbers to and from a time system more to your liking.

If you're really clever or appropriately lazy, you could write a little program to make this RPC call, loop through all transactions, ignore everything outside of the start and end epochs, ignore everything to non-vote addresses, and summarize the results.

Avoid Poll Stuffing

Suppose your author has frustrated a big batch of Zelda fans with the "Sheik" comment and a thousand riled-up people stuff your poll with tiny transactions for anything but the Zelda option. Are they legitimate votes? That's up to you to decide, but wading through all of those transactions to figure it out could be a hassle.

Could you run a poll in a way that reduces the possibility of fraud or ballot stuffing? Absolutely yes, if you know the people you want to vote in your poll beforehand.

Suppose you have 10 friends wildly invested in your blackberry lemonade farm lifestyle. Suppose also, for the sake of math, you want to give them each 10 votes, and they can divide them up among these options with as much precision as they desire. 10 people, 10 votes.

Have each of your friends provide you with a fresh address. Then mine, transfer, generate, or otherwise obtain 101 Dogecoin yourself. In one transaction, send 10 Dogecoin each to each of your 10 friends[16]. Tell your friends that their votes *must* come from the 10 Dogecoin they've each received, so when they craft their transactions either do so explicitly, use an otherwise empty wallet, or use a mechanism such as `lockunspent` (see Protect Unspent Transaction Outputs, pp. 184) to ensure the funds come from the right place.

Then open the poll as normal.

When you count votes, use `listreceivedbyaddress` and the same `blocktime` approach to validate the timing of the votes, but look also at the *source* of funds for each transaction. A vote is only valid *if* the received coins spent in that transaction came from the transaction you used to seed each friend's wallet. Any other transaction is an invalid vote, so ignore it.

This approach also has the nice property of vote weighting, which gives voters more say than the majority-rules approach.

Understand the Risks

Some readers may have noticed that some sets of proposed names don't have the same number of gender-specific names as others. If that bothers some people, they can choose different names for their hypothetical goats, because goats **do not** care about their names[17] and are unlikely to respond when called by any name, let

---

[16] Remember there's a very small transaction fee, so use the 1 extra to pay that.

[17] Allow this author to *assure* you.

alone those you've carefully selected.

In that case, adherence to a theme is more meaningful to the goat herder than to the animals, so let people enjoy themselves where they can!

If you perform this kind of transaction on the public Dogecoin blockchain, remember that the coins have to come from *somewhere*. Anyone who can look at the blockchain and tie votes to your goats has an increased possibility to de-anonymize anyone who participated in the voting, because there's a real world activity associated with multiple public transactions. A voting system conducted in public, with public transactions, with published destination addresses and run for a specific time can give away a lot of details a clever sleuth can use to figure out a lot of information you or your voters might not want to share.

Performing these actions on the testnet or a private network can reduce this risk.

If you use an existing wallet for these addresses, you'll make your life more difficult in counting votes/transactions. Furthermore, you run the risk of mingling koinu used in voting with koinu used in other purposes, which could also de-anonymize you.

What Can You Do With This?

You don't have to limit this to voting for hypothetical goat names. You can use this technique to select between multiple alternatives. With the technique of giving voters Dogecoin to identify them as voters and the ability to split those coins between alternatives, voters can apply stronger weight to some candidates while expressing lesser but still positive preference for others.

If you're willing to put in the effort, you can always consider setting up a *private* Dogecoin (or other) blockchain for voting purposes. This would allow you to control the entire network and all the transactions, providing more privacy and increasing your control over the voting process. Running (or modifying) the Dogecoin software stack is a lot of extra code and complexity that you wouldn't necessarily need in this scenario; it's probably smarter to run a voting-specific blockchain for this if you find yourself polling more than once.

With that said, if you want to get something up and going quickly, you could do a lot worse. Software that exists that you know how to use well is better than alternatives you don't understand or which don't exist.

Tip #58 Protect Unspent Transaction Outputs

Dogecoins are *fungible* in the sense that 1 Doge always equals 1 Doge, regardless of whether it came from a coinbase five years or five seconds ago. When you send Doge in a transaction, you or your wallet can look through all of your unspent transaction inputs to find enough Doge to satisfy your transaction. Any input is as good as any other input.

In practice, you might feel differently about one transaction versus another, in the same way you might have a soft spot for that USD $1 where your best friend drew a mustache on George Washington or that 1989 final printing Canadian dollar that you hope is worth a couple of hundred now because of its scarcity.

While you can always craft transactions by hand if you're into that kind of thing, letting your wallet software move your Doge around might be like letting your pre-teen niece rummage through your cash jar to find just enough money to buy the Minecraft gift card her parents told her she could buy if she earned the cash. If you suddenly can't find Queen Elizabeth II on a tugboat, you'll regret letting your money out of your sight.

Suppose you're super proud of the first work-for-hire gig you did that landed you a cool 200 Dogecoin in a transaction in 2016, and you'd like to do the virtual equivalent of framing it and hanging it on your wall. You can do this in a couple of ways; for now, focus on making that transaction input *unspendable*.

Lock (and Unlock) Unspent Transactions

The RPC command `lockunspent` allows you to tell the Core wallet to treat one or more outputs in a transaction as unspendable. Whenever you attempt to send Dogecoin to an address, the Core will ignore the unspent inputs in the locked transaction.

Suppose you're looking at the second transaction in block 4,012,441[18]. This transaction has two outputs. The first has index 0 and sends 140,000 Doge to an address starting with DLfEFh. The second has index 1 and sends 281,628.30 Doge to an address starting with AC8Q9Z. The entire transaction starts with 5c7561.

If you control the first address and don't want to spend those 140,000 Doge in some random transaction, use `lockunspent` to keep the Core away from those

---

[18]Look this up in a block explorer such as dogechain.info to follow along!

coins when it rummages around in your wallet.

```
$ dogecoin-cli lockunspent false \
  "[{\"txid\": \"5c7561...\", \"vout\": 0}]"
true
```

Be very cautious; this is tricky to get right. The first argument is a boolean true or false value. If you pass true, the command will **unlock** the specified transactions. To lock, use a value of false. Yes, this seems backwards.

The second parameter is optional; it's the serialized form of a JSON array of objects where they keys are the transaction ids and the indices of the outputs in those transactions. Check and double-check your quoting, especially if you use 'dogecoin-cli' to perform this command from the command line.

After you've sent this command, immediately use listlockunspent to see if the transactions you intended to lock are actually locked:

```
$ dogecoin-cli listlockunspent
[
  {
    "txid": "5c7561...",
    "vout": 0
  }
]
```

If you've locked multiple transaction outputs, you'll see the entire list.

To *unlock* a single transaction output, provide the transaction id and output index, changing the boolean argument to true.

Always List Your Locks

Get in the habit of using listlockunspent after *every* change you make to locked transactions. If you send the command lockunspent true without any transaction output identifiers, you will unlock *all* locked transactions.

Arguably the implementation of this command is misleading, though your author admits he had to read the Core source code and think really hard to figure out how it works.

These locks will last until you send an unlock command or your node restarts.

185

Understand the Risks

These locks are effective in small, specific circumstances but they're not security. They won't protect you from losing your keys or having someone else gain control of your *wallet.dat* file. Anyone who can send RPC commands to your node can list any locked transaction outputs and also unlock those outputs.

If you share a wallet between multiple nodes, locking transaction outputs on one has no effect on the others. These locks are stored in memory in the Core only as long as the Core is running. When it shuts down, the locks vanish.

In other words, if you rely on this mechanism to protect a transaction output for whatever reason, *verify* the locks you want active are in fact present *before* starting any transaction.

What Can You Do With This?

This approach offers minor protection, as it's limited to a single node implementation and it doesn't survive node restarts. While future versions of the Core may make this guarantee stronger, you have a couple of other options to protect transaction outputs.

For example, if you use the Action Launcher approach (see Add an Action Launcher, pp. 87), you can write a script that, on every new block detected, makes an RPC call to lock outputs from a list. If you're clever, you could even make this check a PID or log file or flag to keep it from running more than once. When your Core restarts and begins processing blocks, it'll launch your launcher.

Of course, that doesn't protect against a potential race condition, where you might create a new transaction *before* the script has a chance to run. A stronger solution is to isolate the wallet address(es) into a separate wallet not visible to the core where you want to lock these... but that's a discussion for a separate tip!

  Tip #59   Manage Big Transactions Gently

The very thing that makes Dogecoin work as a globally-distributed, permissionless network to send and receive transactions also makes it complicated: miners and nodes will do what they will do when they receive, evaluate, confirm, and retransmit transactions. No single central authority exists that can say "This transaction moving a million Doge is more important than this transaction moving 1 Doge"; it's all about what the network as a whole decides to do.

You can use this information to be a good network participant.

Transactions and Blocks

The blockchain underlying Dogecoin organizes itself in terms of *blocks*. A block is a big batch of transactions bundled together and mined about every 60 seconds. Miners bundle pending transactions into blocks and, when everything verifies correctly, issue a new block for everyone else to analyze and store in the blockchain permanently.

A Dogecoin block is 1 MB in size. That's enough space for 2400 average-sized transactions. The fuller the block, the more efficient the network is. However, the fuller the blocks over a period of time, the more chance that any one transaction will be delayed to a new block. With a mining time of about 60 seconds, every delay of one block can add one minute to confirmation time.

If you are sending a lot of transactions *or* you're sending especially large transactions, there's a chance your transactions could fill up the entire block. In this case, the kind thing to do is to ask the network to de-prioritize your transactions to let other transactions through first.

For example, if you're consolidating a bunch of Dogecoin from a couple of hundred addresses into a couple of new addressees, your transaction will have to refer to every unspent output of those consolidated addresses as well as the new addresses. That could be a large transaction. Ask yourself if you have to finish this *right now* or if it can happen when the network has enough capacity your transaction won't bump others.

Transaction Costs

In the current stable Dogecoin Core release (1.14.9), the recommended transaction fee is 0.01 Dogecoin per kilobyte of transaction size. Unless you create a transaction by hand or change this value yourself via the Core, this is what you'll offer to pay miners to include your transaction in a block. The more core nodes that run code that accepts this transaction fee, the more likely the network will accept your transaction request.

Because this has been the recommended default since Dogecoin Core 1.14.5, this is a good number to use.

To be a good network neighbor, you want to allow everyone else to use the default but you want to pay a little bit less so that nodes will prefer other transactions when filling up a block. If there's space left over, they can pick up your transaction. Pick a value slightly below 0.01 Doge per kilobyte; perhaps 0.0098 Doge per kilobyte. Custom Transaction Fee Configuration, pp. 188 shows an example.

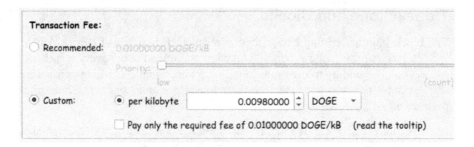

Figure 7.2: Custom Transaction Fee Configuration

With this value selected, nodes which prioritize transactions based on transaction fees will prefer default transaction fee transactions to yours. If the current block is empty, your transaction can go through soon. If there's any kind of congestion, your transaction will wait until everything else clears and there's space remaining.

Understand the Risks

Changing transaction fee defaults always provides the risk of changing how the rest of the network sees your transaction. Set this value too high and you may end up overpaying for a place in a block you'd have reached anyhow. Set this value too low and your transaction might never get processed into a block. Because you don't necessarily know how many other transactions will be pending when the network receives yours, there's always a guessing game involved.

The default should work well in most situations. This approach works best when you're not in a hurry and you want to let other transactions settle more quickly. You won't find yourself doing this often, but when you do, the rest of the network will thank you for it.

  Tip #60   Act on Confirmed Transactions

The Dogecoin network allows people with competing interests to work together through consensus and validation. Truth is what the network decides it is. For example, if 99% of nodes believe that a transaction fee of 0.01 Doge per kilobyte is acceptable, that's what happens, even if 1% of nodes want miners to get a fee of 10 Doge per kilobyte.

Similarly, history is what the network validates. The more descendants any block has, the more you can rely on it and its transactions. While it's *possible* for the network to replay work and redo history from *any* block, it gets more and more difficult, costly, and unlikely with every subsequent block mind. In practice, this means that the blockchain is immutable (it'll never change).

If you're receiving payments, the more blocks that come after any block containing one of your transactions, the more confident you can be that the transaction is permanent. If you're selling cannoli from a pop-up cart in the park, you may care about one confirmation. If you're selling a Banksy painting of a vendor selling cannoli from a pop-up cart in the park, you may want a few more confirmations.

How can you take advantage of this consensus? How can you determine your own confidence in any given block?

Stateful Block Confirmation

The tips Take Actions on New Blocks, pp. 83 and Add an Action Launcher, pp. 87 describe how to react when a running node detects a new block. Finding a transaction in a block is easy enough, if you know the addresses to which you'll receive transactions *and* if you can detect which of those transactions are inputs instead of outputs (see Identify Input Transactions, pp. 172).

With all of those pieces available, to assemble them to do what you want:

- detect confirmed blocks

- ... which contain input transactions to your address(es)

- ... and wait for *n* more blocks to be confirmed

There's only one catch; the example code the in block action tip is asynchronous and stateless by design. In other words, it gets launched as a separate process for every new block. It keeps nothing in memory—including the number of blocks to wait to consider the transaction irreversible.

There are two ways to accomplish this goal. One, make a persistent process that gets called from the block action code. It will have to remember that, if there's an incoming transaction in block 582 and you want 8 confirmations before you hand over a beautiful dog painting, you need to see block 600 confirmed.

The second option is easier: store the data somewhere. Here's a Ruby[19] script that takes a block height and block hash as input, looks for interesting transactions, and stores data in a simple YAML[20] file:

---

[19]Why Ruby? You can use any language you like. Why *not* Ruby?

[20]Why YAML? Part of it was designed in your author's living room. Only the good parts, of course.

```ruby
require 'yaml'

NUM_CONFIRMATIONS = 8
FILENAME          = 'transactions.yaml'

def main(block_height, block_hash)
    block_height = block_height.to_i
    transactions = File.exists?(FILENAME)
                   ? YAML.load_file(FILENAME)
                   : {}

    # find transactions in this block
    inputs = find_input_transactions_for_block block_hash

    # add any transactions found to the tracker
    if inputs.length > 0
        transactions[block_height + NUM_CONFIRMATIONS] = inputs
    end

    # notify any awaited confirmations at this height
    if transactions.key?(block_height)
        transactions[block_height].each do |tx|
            notify tx
        end

        transactions.delete(block_height)
    end

    # write out the transaction file for the next run
    File.open(FILENAME, 'w') do |file|
        YAML.dump(transactions, file)
    end
end

def find_input_transactions_for_block(hash)
    ...
end

def notify(transaction)
    ...
end

if $PROGRAM_NAME == __FILE__
    main(*ARGV)
end
```

The most interesting code here is the logic used to store and retrieve state. Change
the constants at the start of the file to require more or fewer confirmations. 8
confirmations will take somewhere around 8 minutes; too long to sell a cannolo
and not long enough to sell a herd of rambunctious goats. The YAML file can be

in the current directory or elsewhere, named whatever you like.

This code kicks off some code elsewhere (written elsewhere in this book, though not necessarily in Ruby) to find interesting transactions in the block. The method called here, `find_input_transactions_for_block`, should return a data structure or object appropriate for the notification. This could include the sender and receiver addresses, the transaction amount, any transaction label, and the transaction date.

Similarly, the `notify` method should take this object and perform any notification such as described in Add Desktop Notifications, pp. 89 or anything else.

Be sure to do two things in this code. First, add transactions to and remove them from the `transactions` data structure at the appropriate time. Second, write out the file when the process finishes, so that the next invocation can do what it needs to do.

Understand the Risks

If you want to *rely* on this code for telling you when it's safe to release some good or service to someone who paid you, write and run the code as defensively as possible. Be careful to ensure that you don't miss blocks (if the program crashes, fails to run, or otherwise misbehaves). Notifications get sent once per block, and if you miss processing one, you'll have to catch up deliberately.

To make this code more robust, you could store the height of the block you most recently processed, and run back through previous blocks to catch up.

Also be sure to save your file or persist your data with the appropriate level of caution. Storing a plain-text file on your own hard drive allows anyone who understands your data format or who can get at your file to delete, edit, or otherwise modify your data. Assuming a computer will always get things right and never surprise you is a good way to experience unpleasant surprises when the computer gets things wrong.

Finally, make sure that your notification mechanism is reliable. Messages will pop up on your screen, but if you look away or accidentally click it or aren't sitting in front of the computer all day, you might not notice something happened. There's no substitute for human oversight, especially when it comes to dealing with delicious pastry, guerilla art, or fuzzy animals. Guard your valuables appropriately.

Tip #61   Index All Transactions

Dogecoin Core does several things. One, it acts as a client for the network. Two, it acts as an interface to help people decode what's on the network. Three, it acts as a wallet. All three of those things work together, but they don't all work together in the same way you might expect.

For example, if you use RPC calls like `getrawtransaction` to inspect transactions that involved your wallet, you might not realize that, by default, the Core only lets you inspect transactions that involve your wallet. Visit any blockchain explorer[21], pick any transaction ID, then enter it into the Core GUI console and you'll probably get the error message:

```
No such mempool or blockchain transaction.
Use gettransaction for wallet transactions.
```

This may be fine if you're only interested in your own transactions, but if you're writing your own block explorer (Write a JSON Block Explorer, pp. 54) or otherwise examining the network and blockchain as a whole, you need something else.

Enable Transaction Indexing

The `txindex` configuration option tells the Core to examine all blocks and create an index of all transaction IDs, even if they're not connected to any of your wallet addresses. Now any RPC calls that look outside what's reachable from your wallet will work. Add this option to your command-line or your *dogecoin.conf* configuration file.

Understand the Risks

The Core doesn't do well switching back and forth between indexed and non-indexed transactions, so pick one mode or the other and stick with it. Use the configuration file to make your choice permanent.

Indexing all transactions will take time and disk space, so if you're running Core on a small or slow device, consider the tradeoffs carefully. This works better with a full node.

---

[21] Such as https://dogechain.info/.

Tip #62 Test First, Safely

Is there any worse feeling in the world than realizing that your fingers slipped and did the exact thing you didn't want them to do? There are only two solutions to this. First, be perfect always. Second, test what you're about to do, and then do it again exactly as you did it, if you did it correctly.

That's really abstract. Let's make it concrete: before you do something really interesting or risky or new with Dogecoin on the main network, it's smart to try that thing on the test network.

Main Network? Test Network?

There are, in fact, *multiple* Dogecoin networks[22]. You use the main network by default if you don't know there are multiple networks. On this main network, every Dogecoin has real-world value, thousands of nodes collaborate to share history and verify transactions, and lots of miners toil away against each other for the chance to earn 10k Dogecoin every minute or so.

On the test network, the money is funny and there are fewer miners and the stakes are low. Otherwise it works pretty much the same. That's what makes it a good playground.

The Dogecoin Core has a configuration option `testnet`. If you launch the Core with that option (or put it in your *dogecoin.conf* file), the Core will connect to the test network instead of the main network. You can see this in the Qt GUI because the program's title bar include `[testnet]`. Testnet uses ports 44556 instead of the standard main network port of 22556.

Finally, you should see that the testnet mode will download blocks into a *testnet3/* subdirectory of your data directory. Because the networks are different, they don't use the same blocks[23].

What Can You Do With This?

Testnet is a good place to experiment with interesting transactions. Suppose you're hand-crafting something interesting with Dogecoin opcodes, or you're trying to

---

[22]Not including all the pale imitations that exist only to part innocent people from their hard-earned money, most of which are Ethereum tokens.

[23]Hopefully it's clear why; even though the code is the same, the networks have diverged from the start, so the data is all different.

figure out how time-locked transactions work. Rather than risk your actual valuable Dogecoin, load up a testnet wallet and see what happens.

Alternately, if you're working on Core and want to experiment with networking or other code that's intrinsically important to the network, running a testnet node or two with the new code can give you a lot of insight into what's happening and why and whether it's worth making the change to nodes in the main network.

Understand the Risks

Again, the main network and the test network are *different* networks. Dogecoin mined, sent, received, transacted, and/or traded on one aren't valid in the other. If you attempt to send coins from mainnet to testnet, you're likely to lose them. *Do not share a wallet* between mainnet and testnet!

For maximum safety, if you do use testnet, use a different data directory, program launcher, or even a different computer/virtual machine to increase the safety and separation of the code and data and your wallet.

  # Tip #63   Act on Wallet Transactions

Several other tips have led up to this point. If you run a Core node and want to do something on every confirmed block (Take Actions on New Blocks, pp. 83) or transactions (Act on Confirmed Transactions, pp. 188), you may find yourself asking "But what about transactions to addresses I care about?"

Fortunately, you can get even more specific, especially if you have a list of addresses you *really* care about.

Wallet Transaction Action

The `walletnotify` configuration option works like `blocknotify`. Add to your *dogecoin.conf* a line that provides the path to a program to launch. For example, if you wrote a shell script called *walletnotify.sh*, you might add the configuration option:

```
walletnotify=/home/myuser/bin/walletnotify.sh %s
```

Provide the full path to the file, make sure you can execute it (this means `chmod +x` on Linux and other Unix-like systems), and note that the `%s` option to the configuration tells the Dogecoin Core to provide the transaction hash as the only argument to the program when it runs.

As with all of the other launched notifiers, each invocation is independent and stateless. Remember: if you care about doing something once and only once for each independent transaction, it's up to you do track that you've seen the transaction, started to do the thing, and finished doing the thing.

If you're popping up a notification on your desktop or asking your smart speaker to play Monty Burns saying "Excellent" on every transaction, you may not care about duplicates. If you're allowing monthly subscription access to your secret website full of delicious cake recipes, accidentally duplicating transactions will cost you!

A good design might be:

- create a transaction notify script

- insert a work item to *process* that transaction in a persistent queue (if it doesn't exist)

- process the transaction (make the RPC calls to get time, amount, source, et cetera)

- mark the transaction as processed

By separating the *recording* of the transaction from its *processing*, you can ensure that taking whatever action you intend to take succeeds or fails all at once in a separate system, independent of however the Core launched your notify script.

Is This Better Than Confirmed Block Watching?

How is this different from acting on block confirmation? That depends what you want to do. The other tip demonstrates how to set a countdown from the time the network accepted a block containing a transaction you cared about until a point in the future when you consider that transaction fully settled and irrevocable. This tip is *compatible* with that approach, but different in the sense that all it does is look at transactions, not blocks.

Sometimes you want both: detect that a transaction occurred *to* a wallet under your purview, figure out the height of the block where you consider that transaction complete in the blockchain, then wait for the block at that height to be confirmed.

What if the Core could also include the block height in the arguments to the notify script? As of Dogecoin Core 1.14.8, use the %i option in your `walletnotify` configuration to include this height.

Understand the Risks

In addition to every risk listed in the other tips, be aware that reindexing the blockchain, whether directly with `rescan` or implicitly with an address import, will cause re-notification for all of the transactions associated with all of the addresses in your wallet, watch-only or fully-keyed. This can be desirable, if you're importing addresses and want to take action on them.

This can also be an unpleasant surprise, if you find yourself processing transactions over and over again.

Be mindful of what you want to achieve and consider a persistent transaction processing mechanism so that you can avoid repeating work, especially if that work has side effects.

  Tip #64   Queue Actions on Transactions

The previous tip (Act on Wallet Transactions, pp. 194) proposed an architecture for turning incoming Dogecoin transactions into concrete actions in other systems. This is important for building things such as a Dogecoin-powered arcade (Manage a Dogecoin Arcade, pp. 301), but it's not limited to making lights flash and sirens go off. Almost any payment system can use this approach.

Assuming you have a wallet with a few addresses, a Core node, and notifications set up for wallet transactions, what happens next? You need some glue between "something happened in my wallet" and "do something about it".

Inside a Transaction

What is a wallet transaction anyhow? Your wallet contains an address that was used as an output in a transaction validated by the network. If the network continues to accept that transaction and build on it, you can spend those coins in future transactions provided you have access to the private key associated with that address.

However, things aren't as easy as they seem. For example, just because an address is present in a transaction output doesn't mean that the funds are new to that address; you might use a Dogecoin client that pays change back to the same address[24]. Alternately, you might have an address that's used to share in the re-

---

[24]This is generally a bad idea; it's safer and simpler to use a new address for each transaction, but *you can't always control this with all clients.*

wards from a mining pool and receive a coinbase payout that doesn't represent a *customer* transaction.

Finally, depending on your use case, you may want to wait for a certain number of blocks mined *after* yours, to ensure that the transaction is irreversible.

Sometimes all of these circumstances *may* be meaningful to you. You need some sort of system to understand the context of these transactions before you let them trigger an action that's difficult to reverse, such as queueing Spacehog's "In the Meantime" 99 times on your jukebox[25].

This is heady stuff. Let's talk code.

Fetch and Store Transaction

Assume you have a transaction notifier script enabled. It can do *something* every time the Core kicks it off. What will it do?

First, fetch and decode the transaction (see Write a JSON Block Explorer, pp. 54 for one example). You'll want input addresses, output addresses, transaction date and time, block height, and transaction amount at least–perhaps other fields.

Assuming you have a database storing things like wallet addresses (see Rotate Machine Addresses, pp. 307), you can create a new table to store basic transaction information too:

```
CREATE TABLE transaction (
    transaction_id   TEXT NOT NULL PRIMARY KEY,
    block_id         TEXT NOT NULL,
    transaction_time TIMESTAMP WITHOUT TIME ZONE NOT NULL
);

CREATE TABLE transaction_input (
    transaction_id TEXT  REFERENCES transaction(transaction_id),
    from_wallet    TEXT  NOT NULL,
    value          FLOAT NOT NULL
);

CREATE TABLE transaction_output (
    transaction_id TEXT  REFERENCES transaction(transaction_id),
    to_wallet      TEXT  NOT NULL,
    value          FLOAT NOT NULL
);

CREATE INDEX transaction_output_to_wallet_idx
        ON transaction_output (to_wallet);
```

---

[25] It's a banger, but anything 99 times in a row is of a muchness.

You can get fancier than this basic example. For every transaction notification, grab and decode its contents, then populate these tables.

Why store this information persistently? It's useful for debugging and auditing *and* if your node has any interrupted communication, you want to know how to pick up where you left off. If a customer sends you an email or walks up to your service kiosk in person and says "Where's my lemonade? Are you still muddling the blueberries? I sent your Dogecoin 10 minutes ago!" you should be able to examine your list of received transactions and verify what happened[26].

Analyze Transaction

Once you've stored a transaction and saved it to disk (not only in memory), you can process it according to your rules. If you're waiting for 6 or 10 or 100 further blocks mined before you do anything, you can set another notification watcher to move to the next step only at that block height if the transaction is still valid.

Otherwise, check the other conditions: is this a coinbase transaction? Did you send yourself change? Is the transaction amount above the threshold you consider meaningful?

If you're *really* disciplined, you could have multiple mechanisms at play. For example, if you give every new customer a unique address for unique contexts, you could have simultaneously multiple addresses awaiting coins to:

- Unlock the member's area of your website

- Download a whitepaper about Dogecoin transaction handling

- Buy tokens to play LOTR pinball in your ice cream shop/arcade

- Pay for a pint of ice cream in your arcade/ice cream shop

Once you've decided a transaction is both valid and meaningful for your purposes, you can prepare to perform any arbitrary actions in response.

Queue Potential Action

Your author prefers an architecture of registering event handlers invoked by queued transactions, but that's not the only way to design this system. On the other hand, a lot of people know Python, JavaScript, and PHP these days, so the HTTP-style event-based architecture is a good fit.

---

[26]True story: your author pre-sold this book's first PDF to Timothy Stebbing but the purchase unlocking failed because of an error in the notification script. The *debug.log* showed the error and the database table revealed that the transaction had not been recorded.

One easy way to do this is to add tables to register an event handler with an address:

```sql
CREATE TABLE action_type (
    action_type_id SERIAL PRIMARY KEY,
    name           TEXT NOT NULL,
    display_name   TEXT NOT NULL
);

CREATE TYPE action_status AS ENUM (
    'pending',
    'active',
    'expired'
);

CREATE TABLE action (
    action_id         SERIAL PRIMARY KEY,
    created_timestamp TIMESTAMP WITHOUT TIME ZONE NOT NULL
                      DEFAULT CURRENT_TIMESTAMP,
    expires_timestamp TIMESTAMP WITHOUT TIME ZONE NOT NULL
                      DEFAULT CURRENT_TIMESTAMP,
    action_type_id    INTEGER
                      REFERENCES action_type(action_type_id),
    status            action_status NOT NULL,
    to_wallet         TEXT  NOT NULL,
    contact_email     TEXT  NOT NULL DEFAULT '',
    action_args       JSONB NOT NULL
);

CREATE INDEX action_to_wallet_idx ON action (to_wallet);

INSERT INTO action_type (name, display_name)
VALUES
    ('Activate',       'internal action'),
    ('BoredSnape',     'mint a Bored Snape NFT image URL'),
    ('DogeBookAccess', 'grant access to a book'),
    ('PostToDiscord',  'activate a Discord webhook'),
    ('PostToSlack',    'activate a Slack webhook'),
    ('QueueSong',      'queue a song in the jukebox'),
    ('InsertCoin',     'insert coin into an arcade machine'),
```

Take this one step at a time. For every action–every concrete thing you want to *do* when you receive a meaningful transaction–insert a record into the `action_type` table. For every action you want to take for every meaningful wallet address you have, insert a record into the `action` table.

Every action type implies code somewhere else in the system that knows how to do something. This design creates a meaningful interface between "a transaction happened", "the transaction is meaningful", "the transaction triggers one or more actions", and "make the action happen". Modify to your liking.

In practice, this system works well when you treat these actions as invoking webhooks (see Post to Discord, pp. 64, for example).

Understand the Risks

Any time you connect a system to the Internet, you incur the cost of maintaining and protecting that system and borrow the risk of an adversary compromising your system or data. Building network services from Dogecoin transactions includes those risks.

Furthermore, if you're listening for events from a Dogecoin Core or equivalent as part of a payment processing system, you inherit risks from the software and network itself: the network may split, transactions may take a while to process, your node may have a security vulnerability and need to restart, you may run out of disk space, or your network may go offline.

You can recover from all of these, and a good architecture can help, but you don't have anyone you can call up on the phone[27] and ask for help. You're on your own.

That's kind of fun and sobering at the same time.

  Tip #65   Timelock a Vault

Other tips in this book demonstrate that the network will only validate a transaction if the funds are otherwise unspent *and* if the transaction satisfies a mathematical puzzle or lock for its incoming transactions. Traditionally, this puzzle or lock is "prove the person attempting to spend the input transactions has a private key which corresponds with a public key associated with the incoming funds", but the puzzle can be anything (Host a Treasure Hunt, pp. 174).

Dogecoin inherited a complex set of operations from Bitcoin that you can use to create multiple types of puzzles used to unlock funds. These operations are a mini programming language with rich potential[28]. One such operation is the *timelock*, which allows you to create an input transaction unspendable until a specific point in time.

---

[27]Let's pretend it's the 20th century and you want to use a telephone for voice purposes.

[28]...both to perform complex and useful operations *and* to create complicated ways to lose your funds forever.

200

Create a Vault with a Time-Based Lock

BIP-65[29] defined a new transaction operation called `OP_CHECKLOCKTIMEVERIFY`. This operator takes one argument, either the number of seconds since the Unix epoch time (January 1, 1970) or the blockheight.

How do you distinguish between the two values? Dogecoin Core uses a threshold value called `LOCKTIME_THRESHOLD` (you can see it in action in BIP-65, but at the time of current writing, the BIP does not explain this value clearly). As of Dogecoin Core 1.14.9, the value is 500000000, or 500 million. Any value less than 500 million is interpreted as a block height. Any value greater than 500 million is interpreted as epoch seconds. Right now there have been fewer than 5 million blocks mined, so there's room to grow.

You can write this transaction by hand, or you can use an established pattern. Fortunately, Peter Todd, the author of BIP-65, devised a template for this. The pattern (see Decode a Transaction Script, pp. 203) looks like:

```
<expiry time> CHECKLOCKTIMEVERIFY DROP
DUP HASH160 <pubKeyHash> EQUALVERIFY CHECKSIG
```

You must provide two values. First, the expiry time is either a time in Unix epoch seconds or the height of the blockchain after which the transaction can be spent. The second value is the hash of the public key associated with the address containing the funds to be unlocked.

Tools to Create a Locked Vault

Patrick Lodder, Dogecoin Core maintainer, has written software to generate vault data for you. His tool is in a repository called `cltv-vaults`[30]. You'll need a working Node.js runtime with `npm` installed. Use `npm` to install `bitcoinjs-lib` as well.

With the `clvt-vaults` repository downloaded, run:

```
$ node examples/create-vault.js
WIF:            cVAJx...
Public Key:     e226c1a...
Vault height:   31338n
Vault script:   026a7ab17...
Vault address:  2N2ezZvs9yzysMcqq3b5yPCdXQFGRnP4Szw
```

---

[29] See https://github.com/bitcoin/bips/blob/master/bip-0065.mediawiki.

[30] See https://github.com/liberateinc/cltv-vaults.

201

This script will generate a new private/public keypair and a script that you can use to lock funds sent to the vault address until block 31338–but that block has come and gone.

Open the example code in your favorite text editor and look for a line like `const UNLOCK_AT = BigInt(31338)`. Change the number in the parentheses to the height you prefer (after making sure it's sometime in the future). Run the script again, and now you can do two things. First, use the `WIF` value to import into a Dogecoin wallet. This will work with the Core or any other wallet that supports the Wallet Interchange Format (see Interchange Your Wallet Keys, pp. 15).

Second, send the funds you want to lock in the vault to the provided address. **Beware** that if you lose the WIF, you'll lose the private key. Even after the lock expires you won't be able to unlock the transaction and recover the funds.

Tools to Verify the Locked Vault

Once you have that script, verify it does what you think it should do. Copy the "Vault script" value and use the Dogecoin Core RPC mechanism to use the command `decodescript`.

```
decodescript 026a7ab17...

{
  "asm": "31338 OP_CHECKLOCKTIMEVERIFY OP_DROP ... OP_CHECKSIG",
  "type": "nonstandard",
  "p2sh": "A1r3F312ScMR4Caknb8X1PGdtUSJB96Cxo"
}
```

For now, ignore the second large number in the `asm` value. Check that the first value in the script (the contents of `asm`) matches the block height or epoch seconds value, whichever you chose, and that the second item in the script is truly `OP_-CHECKLOCKTIMEVERIFY`.

If and when you're comfortable with this, you can publish your transaction to the network (see Put Funds in Escrow, pp. 206). Then any funds you send to the P2SH address are out of your control until the network reaches the threshold you've set.

Uses of Vaults

BIP-16 suggests several possibilities, such as locking funds in a multi-key scenario where two or more people can unlock the funds right now, and a third is available only after a certain time period.

Another approach lets you deposit funds into a vault for a child and unlock the transaction only when that child reaches 18 years old. Alternately you can stagger payments and ladder the funds: one transaction at 18, one at 19, one at 20, and so forth.

Yet one more idea is to create a vault that can receive public funds the bulk of which can't be spent until a specific time, for example to crowdfund a book, album, board game, or another creative endeavor. With this approach you can publish the entire transaction script and allow other people to verify its intent.

Understand the Risks

Even after the time in the transaction has expired, if you lose the private key and/or the ability to generate an unlocking script, your funds are locked in an unspendable transaction.

Furthermore, be aware that the risk of using block height is that the predictable 55-60 second mining rate may not hold in the future. If mining gets faster and cheaper, the time between blocks might go down. If miners disappear, the time between blocks might go up. Although the network adjusts difficulty to adapt to these situations, the 1 minute per block heuristic is an estimate, not a hard guarantee.

Finally, if you've just celebrated the birth of a new child in your family and want to lock up Dogecoin for 18 or more years, consider everything that happened in the past 18 years and think about what could happen in the future. Even only nine and a half years ago, when Dogecoin was created, it wasn't clear that this joke currency would have a 22 billion dollar market cap and an amazing book with recipes for blueberry lemonade mixed with references to classic video games and goat-based treasure hunts written about it.

Predicting the future is difficult, especially for things that haven't happened yet, so beware of locking things in Vaults (even or especially Vault-Tec vaults)!

  # Tip #66   Decode a Transaction Script

If you do anything more complex with Dogecoin transactions than letting your wallet send funds, be doubly careful that your transaction does no more and no less than what you intend.

Consider this: the central activity of a Dogecoin transaction is to ask nodes and miners to verify that a piece of data from your transaction provided to a little bit of code encoded in the input transaction produces a true value *when executed*. This powerful and dangerous idea gives the blockchain power while asking multiple other computers to run tiny computer programs on your behalf.

Programs often have bugs. If you're creating transactions with any kind of logic in them, double- and triple-check that they do what you think they should, lest you lock up funds so they're unspendable or allow anyone else to spend them.

This is especially true if you use third-party software to help you craft transactions.

Parts of a Script

Script has two parts. First, the script itself comes from the input transaction. This is a lock on funds for which you need the key to spend. The second part is the data the spending transaction provides. In other tips (Host a Treasure Hunt, pp. 174), this could be a password or phrase or other puzzle.

Some scripts, such as timelocking (see Timelock a Vault, pp. 200), are complex. To unlock the transaction, the time or block height had to be after a specified threshold *and* the spending transaction had to come from someone with the right private key.

For a complex script like this, BIP-16[31] defined a way for Bitcoin (and other cryptocurrencies which adopted this scheme) to treat the script as an address which can receive transactions. This scheme is P2SH or Pay to Script Hash.

P2SH in Action

The previous tip about timelocking was inspired by an actual transaction used for Dogecoin developer tips. Core maintainer Patrick Lodder created a P2SH address from a timelocked transaction and published the address and the script for other people to verify[32].

The P2SH address is 9rbpxxkjB9uZbcZzZCPibnoiotGkPoGFEJ, and the raw script (with spaces added for formatting–remove them to decode) is 03 d0 75 41 b1 75 76 a9 14 80 74 43 40 ca 22 90 b6 66 ac 6d 4f c9 8d c1 bb c6 4e 48 c8 88 ac.

Verifying by Decoding

Patrick used custom software to create the transaction and the P2SH address, but gave people everything they need to verify that what he wrote is accurate. *He verified the script himself through alternate means, figuring that, if multiple decoding methods all give the same results, he could trust them–and other people could trust him.*

---

[31] See https://github.com/bitcoin/bips/blob/master/bip-0016.mediawiki.

[32] See https://www.reddit.com/user/patricklodder/comments/v9lvlb/accounting_for_dogecoin_core_development_tips/.

Use the `decodescript` RPC command from either the Dogecoin Core debug console or `dogecoin-cli` with the raw script (03d075...) as a single argument to see the result:

```
{
  "asm": "4290000 OP_CHECKLOCKTIMEVERIFY \
          OP_DROP \
          OP_DUP OP_HASH160 \
          80744340ca2290b666ac6d4fc98dc1bbc64e48c8 \
          OP_EQUALVERIFY OP_CHECKSIG",
  "type": "nonstandard",
  "p2sh": "9rbpxxkjB9uZbcZzZCPibnoiotGkPoGFEJ"
}
```

These three fields tell you a lot: this is a "nonstandard" script, which means (more or less) the address is not the hash of a public key. The `p2sh` field reveals that what Patrick claimed is correct; send funds to this address which can be unlocked by fulfilling this script. Finally, the `asm` field decodes the script itself into a human-friend representation of the opcodes and data used to construct the script.

After block height 4,290,000, the locktime will release. The script will then remove the true value (returned from the previous opcode), then duplicate the next item on the stack (the key provided by the spending transaction), hash the duplicate, verify that the hash matches the one provided in the script, and finally verify that the key provided by the spending transaction matches the expected key and script signature.

All that is left is for anyone to verify that the HASH160 of the key Patrick intends to use matches the provided value.

You can decode any transaction script this way. All you need is a raw transaction, such as the output of the `getrawtransaction` RPC command. However, at the time of this writing, most transactions are very simple payments to the hashes of public keys.

Understand the Risks

What's the risk here? If you're not careful, bugs could ruin your day. Writing your own transactions always carries risk—not only that you'll make a mistake, but you'll inadvertently use a feature or a mechanism that the network won't understand the same way you do.

Use `decodescript` liberally, but don't consider it the be-all and end-all of your research. If you're using a different transaction mechanism you've seen elsewhere, compare the output you get to the script published. Then compare the variables—the data in the script—to the data you intend to use. This is especially true if you use a third-party tool to create the transaction.

If you're determined to do something new and invent your own script, try it multiple times on the testnet first (see Test First, Safely, pp. 193).

  ## Tip #67　Put Funds in Escrow

Suppose you commission an artist to paint a picture of your cute little pupper. This will hang in a place of honor in your office, so it's a large painting. You agree to pay 10,000 Dogecoin. This is a lot of coins, so you're anxious to get your work. The artist has to buy and stretch a canvas, buy paints, do some sketches, and risk carpal tunnel trying to get all of the little eyelashes of your pooch just right.

If things go south, you want to get your money back. If you stiff the artist[33], they'll have trouble selling the painting of your mutt to someone else.

How can you reduce the risk for both of you? Put the money in escrow.

N of M Multisig Scripts

Another tip demonstrated how to send funds to an address that you can't spend until a certain time or block threshold has passed (see Timelock a Vault, pp. 200). Dogecoin inherited another interesting script feature from Bitcoin: the requirement that multiple keys sign a transaction before you can spend it[34].

The OP_CHECKMULTISIG script opcode takes multiple arguments: a list of public keys used to verify the spending transaction signature and a number. This is important: multiple public keys means that you need multiple *private* keys to sign and unlock a transaction. If multiple people use their own public/private keypairs, you can require that more than one–but not every one–unlock the transaction.

For example, if you and your significant other are saving for a trip to Kauai, the garden island, you might each send your Dogecoin change to a special address. When it comes time to buy your plane tickets, you both need to sign the transaction to unlock the funds, thus avoiding a hilarious Gift of the Magi situation. This would be a 2-of-2 multisig script.

When considering how to treat you, your doggo, and your painter fairly, you might prefer a 2-of-3 multisig script, where you and the artist each contribute a public key and a trusted third party provides the third. In the ideal situation, you and the artist each sign the spending transaction and everyone is satisfied. Else, one or the

---

[33] Please never do this.

[34] See https://developer.bitcoin.org/devguide/transactions.html#multisig in the Bitcoin documentation for more details.

other of you convince the escrow agent to sign the transaction without the other party agreeing. You can add arbitrary complexity, as long as you want to wrangle the signatures, and if you define "arbitrary" as "probably about 16"[35].

Use a Multisig Script for Escrow

How do you create this script? Assume you've agreed on an escrow agent. Each of you should now generate a public/private keypair. Keep your private key private and safe.

All three of you should share your public keys. The safest way to do this is for you and the artist to send your keys separately to the escrow agent; that way you can be sure the agent has both keys. Presumably you trust the agent.

Create the Multisig Script

Now the agent can create the script using the `createmultisig` RPC command:

```
createmultisig 2 \
  [ 'your public key', 'artist public key', 'escrow public key' ]
```

The result will be a JSON object with two keys: `address` and `redeemScript`. The escrow agent should send this to both so that you can verify the script (see Decode a Transaction Script, pp. 203). Be sure that your public key is in the right spot when you decode the script. `address` is where to send the 10k Dogecoin you agreed on. When the artist sees that this transaction has been mined, they can begin to paint.

Spend the Multisig Inputs

What happens when something happens? If two of you three parties are satisfied, you can spend the funds. This is where things get tricky. First, you have to sign the transaction *in the order in which your public keys appeared when creating the script*. If the escrow agent created the script with keys in order of Brock, Orpheus, and Rusty[36] then Brock must sign before Orpheus or Rusty *or* Orpheus must sign before Rusty. If you sign in the wrong order, you won't unlock the funds. This applies for every *n* of *m* script.

Second, someone (ideally the escrow agent) has to create a raw, unsigned transaction, then get two of the signers to sign the transaction in progress to create a fully signed transaction.

---

[35]Technically, the number allowed may be more than a hundred, but is that practical?

[36]On second thought, Rusty may be the *worst* choice for an escrow agent.

Use the `createrawtransaction` RPC command to start this process; provide the `txid` of the funding transaction(s), the `vout` index of the input in the transaction you want to spend, the `scriptPubKey` of the funding transaction(s), and the `redeemScript` from the multisig script, as well as the destination address of the spent funds and the amount of spent funds. This operation will return a raw transaction (called `hex`) you can now use as the first input to the `signrawtransaction` RPC command. The second input is the JSON-encoded text of the raw transaction. The third input is the private key of the first signer, in this case either Brock or Orpheus.

Then do the same thing again, with the new `hex` from the previous step, and the next private key, either Orpheus (if Brock signed) or Rusty (if Orpheus signed).

Finally, use `sendrawtransaction` to send the results to the Dogecoin network. Nodes will validate and transmit it and miners will mine it into a block. You'll have a painting of your faithful companion in your office and an artist will have a few thousand Dogecoin in their wallet.

Wait, What About the Escrow Agent?

This is a lot of work. You should probably also pay the escrow agent. Figure out something fair. Remember to account for transaction fees, too.

Of course the *spending* transaction can have arbitrary complexity, like sending 10,000 Dogecoin to the artist and 100 to the agent (if you funded the p2sh address with 10,100 Dogecoin).

Timelock a Multisig Script

If the spending transaction can be arbitrarily complex, what about the script itself? Payment scripts can also be arbitrarily complex. You can write any valid script code and submit it to the network[37].

If you want to prevent anyone from spending the funds for a period of time–after all, painting a good picture of your very good boi or girl will take at least a month–you can put `OP_CHECKTIMELOCKVERIFY` commands at the *start* of the script, so that the block height or epoch time will have to pass before anyone can even *attempt* to validate multiple signatures.

Other script options are possible too, though the more non-standard your script, the fewer nodes may relay it by default.

---

[37]Not all nodes will retransmit non-standard scripts, but some will.

Understand the Risks

What can go wrong here?

Any complex transaction runs the risk of human misunderstanding, and that could mean putting your hard-earned Dogecoin at risk, especially of being unreachable.

Escrow exists so that a subset of people can complete a transaction without everyone agreeing to it. If you are worried about dealing with untrustworthy or unreasonable people and find yourself making ever more complex transactions, step back and think about the kind of stress you want to experience in your life. A simple transaction is easy to understand and easy to execute. A complex transaction takes more effort, requiring more steps. More steps means more opportunities to get things wrong.

Is it worth creating transactions by hand like this? Balance your risk tolerance with your desire for control. Relying on software someone else created to do the right thing introduces some risk, but the convenience of well-tested and repeatable code is important. **Do not** blindly copy the code or instructions here without testing, verification, and validation–there's a reason the "how to generate keypairs" section is left as an exercise for the reader.

  Tip #68   Crowdfund Crowd Fun

While the default assumption of a transaction-based action dispatch system (see Queue Actions on Transactions, pp. 196) maps a single transaction to a single event (play an arcade game, buy a muffin, listen to Copeland James's "Darling We've Got Time"[38]), assumptions are made to be broken.

You might need a different model, if you're holding a naming contest (Vote on Goats, pp. 179), subscribing to a private service for a period of time, or even running a fundraiser for something awesome. Crowdfunding is a great example; something cool will happen only when transaction amounts sent to an address or addresses meet or exceed a specific threshold!

While re-using a wallet address can be risky (more on that later), there are some advantages. Be mindful of the pros and cons.

---

[38] See https://www.youtube.com/watch?v=CMgVXyRJ_Lo.

Threshold-Based Action Dispatch

Suppose you use the transaction/action dispatch pattern from the transaction action queuing system. When a transaction occurs, your Core node notices it and performs a notification action. Your notification system records the transaction information and dispatches to a custom action.

That action can be anything. For example, an `action_type` of `Crowdfund` could keep a running total of all transaction amounts. For each new deposit, it could calculate the amount remaining and do several things:

- Update a status page, showing number of funders, amount funded, et cetera

- Reveal new rewards for reaching stretch goals

- Unlock something cool upon reaching every threshold

Coding a Crowdfunding Action

What does this code look like in practice? The action dispatcher needs to provide some information to every invocation, starting with the appropriate receiving address, address received amount, timestamp, and possibly number of blocks of confirmation). The action also needs to be able to associate each receiving address with a unique campaign[39].

You might write code like this with Perl or any other language:

```perl
use Modern::Perl '2024';
use Object::Pad;

class Crowdfund {
    field $projects;

    method action( $transaction, $current_height ) {
        my $address = $transaction->receiving_address;
        my $project = $self->find_project( $address );

        $project->add_tx( $transaction, $current_height );
    }

    method find_project( $address ) {
        return $projects->find_project_by_address( $address );
    }
}
```

---

[39] Why a unique campaign? Suppose you have multiple campaigns, each with different goals. Consider this flexibility.

210

```
class Project {
    field $confirmation_threshold;
    field $funding_threshold;
    field $current_amount;
    field $funder_count;

    method add_tx( $transaction, $current_height ) {
        my $tx_height = $transaction->height;

        # tx is too young to be confirmed
        return if $tx_height + $confirmation_threshold
                > $current_height;

        $current_amount += $transaction->amount;
        $funder_count++;

        $self->fund
            if $current_amount >= $funding_threshold;

        $self->update_funding_status;
        $self->check_stretch_goals;
    }
}
```

This is skeleton code of an object-oriented design. The example doesn't deal with data persistence, but a decent programmer should be able to manage that without too much trouble.

The transaction action dispatcher invokes the Crowdfund object's action method, which uses data from the $transaction object to look up project information, delegating that approach to some $projects object reachable from a Crowdfund instance. Assuming a project exists, the Project class handles other, more interesting behavior.

What does Project need to do? It has plenty of data, including the number of confirmations required before it considers a transaction confirmed, the amount of funding required to unlock the project at all, the current amount of funding, and the number of funders.

Find the Bug!

There may be at least one subtle bug in Project's add_tx method. Can you find it? Hint: is there a one-to-one relationship between a transaction and a funder?

The `add_tx` method code checks the transaction age against the project's specific threshold, adds the transaction amount, increments the number of funders, and then decides what to do next.

What happens when the project exceeds its first funding threshold? The `fund` method, left as an exercise for readers, should idempotently[40] do something cool. The `update_funding_status` method, also left unimplemented here, can send emails and/or update web pages (or data sources backing web pages) or anything else. Finally, the `check_stretch_goals` method can be part of `fund` (maybe it should be), but treating it separately identifies that there may be unique behavior modeled by `Project` that isn't the same as funding.

There's nothing special about this code; it's straightforward and even boring. That's the point! If you think about the *mechanics* of associating a Dogecoin transaction with a project in a sensible way, you can ignore most of the *mechanics* of payment. All you have to know is when funds come in at a confirmation level you're comfortable with.

Even *that* is configurable here.

Understand the Risks

You can run this one of a couple of ways. If you generate a new address for each potential funder, you need some way of associating a project with an address. That's not difficult!

You also *probably* want to know who's funding your project, especially if they get specific rewards at specific thresholds. In that case, one of your largest security risks is exposing your funder list and, transitively, their addresses and transaction histories. To *reduce* this exposure, remove any association between an email address or other contact information and a transaction address when you confirm the transaction. This limits the audit trail of your system and requires you to generate a new address for any subsequent transactions. That increases your security.

Think of it this way: an address that hasn't received anything *yet* exposes no previous transaction history of the funder, because there's nothing to see yet. Of course, a partially funded transaction could expose the funder's transaction history.

Either way, consider the security of your system and the privacy of your funders as well as your own.

---

[40]This is a fancy programmer word that means "Do it only once. If you try to do it again, don't do anything again."

Tip #69   Referee a Raffle

Anywhere someone has to hand money to someone else is an opportunity for someone to transfer Dogecoin to someone else. That's the *currency* part of the word *cryptocurrency*. The more places and situations where you can spend friendly dog money, the more useful it becomes.

Even better, the more situations where *programmable* friendly dog money can be appropriate, the more *interesting* Dogecoin becomes. The immutable, public ledger offers opportunities to build transparency and fairness into situations where other people might have to take your word for trustworthiness.

Consider a raffle as example: a lottery where people buy tickets and one or more drawings awards prized based on the purchased tickets. How do you know the drawing is fair? With cryptocurrency, you can make the drawing itself 100% provably fair and reproducible. Here's how.

Prove Your Work

Outside of a blockchain, you can sell a raffle ticket, tear it in half, and give one half to the buyer. You keep the other. When it's time to draw a winner, put all of the halves in a container, close your eyes, and pull out the winning ticket. Call the number, and let the person with the other half show you their ticket and verify that the two halves match.

How do you do this with a blockchain? There might not be physical tickets. You might not all be in the same physical location. You might not even know who the (anonymous) winner is.

That's okay; you can still make this system work. The key is to make the *drawing* itself provably fair and transparent.

Pick a Winner

Suppose you have a hundred tickets available. Choose a random number. Publish that number along with a list of 100 unused Dogecoin receiving addresses (see Derive More Addresses, pp. 318). The first person to send a transaction to each address has successfully bought a ticket. When you've sold all 100 tickets, you have a list of 100 addresses as well as the 100 transactions. List these transactions all in order in which you received them–the first is transaction 0, the second 1, and so forth. Write down the index of each of these transactions in their blocks–this time starting from 1.

Now choose another random number. It doesn't matter what it is, as long as it's random. Change every transaction id into a number. Multiply each transaction number by its index in its block. Multiply the result by your random number. Now add all of those numbers together. Finally take the modulus (see Roll Over Your Odometer, pp. 5) of that result and 100 (or just take the last two digits of that number–but if you sell a different number of tickets, use the number you sold as the modulus). The final number there is the number of the winning transaction.

Contact the winner somehow[41].

Given that you've published the random seed and the list of transaction addresses publicly, anyone can verify the list of transactions and their order to reproduce your results. You can reproduce the results–even write a program to do this for you or anyone else. As long as you publish the algorithm you're using (assign each transaction an index number, starting with the earliest first, then the earliest in the block if two transactions get mined in the same block, starting from the first random seed, modulus by the number of transactions), you've achieved your goal of transparency.

Pick Multiple Winners

What if you have multiple prizes? For each item you're awarding, remove the winning transaction from the list and reduce the modulus by one. Repeat this process and continue until you've awarded everything. The same rules apply: make the algorithm and all of its input data available to anyone who wants to check your work.

Can This Be Gamed?

If all of this information is publicly available, can someone game the system? That depends on their ability to *predict* the index of the transaction that will win any specific prize–or, in the case of multiple prizes, the indices of all of the winning transactions in order.

Given that the inputs to the algorithm are your random seed and all of the relevant transaction ids, this is *probably* good security. If you publish your seed before anyone sends a transaction, people can trust that you haven't pre-selected a winner by choosing multiple seeds that send the prizes where you want them to go. That's good.

However, by publishing the seed and all the transaction addresses, you've also given people the ability to monitor the blockchain for eligible transactions. The person who sends the final transaction may have the ability to influence the final

---

[41] This may be the tricky part.

214

transaction id[42]) to influence the final result. That's why the transaction id gets multiplied against the transaction's index in its block: it's more difficult to influence *that* number.

Difficult but impossible? It's not clear yet. This is why good cryptography validated by experienced cryptographers is so important. Any mistake you make that leaks information or gives anyone an advantage over anyone else is a potential attack vector.

Weigh Sold Tickets

If you've ever bought a raffle ticket before, you've probably noticed that you can buy more than one ticket a time. Perhaps one ticket costs $5 but you can buy 5 tickets for $20. This is great–you get more chances to win and the raffle organizers raise more money.

This complicates your job as a referee, because you have to account for the *number* of tickets purchased, not just the *number* of transactions. Of course, this is probably what you want anyway, because you've already invalidated transactions that didn't go through, were refunded, or didn't meet the minimum entry fee.

The computer science/algorithms solution to this problem is to use a technique called "weighted random" selection. Another approach is simpler: add one entry for each ticket of each transaction in order. For example, if each ticket costs 10 Dogecoin and the first transaction was for 30 Dogecoin, add that input address to the list of candidates three times.

Understand the Risks

While this tip is *nominally* about how to hold a fair and transparent raffle with virtual payments, it's also about how to *design* a fair and transparent mechanism where everyone can verify the results. When designing a system like this, you might inadvertently make a loophole that a clever attacker can exploit. Think about this system and how it works and how you yourself might exploit it. If you decide to use the system described here, buy a couple of sneaky friends dinner or a delicious beverage and brainstorm ways to cheat. If you can't think of any, that doesn't mean you're okay–but it's a start.

Another risk unique to a virtual, distributed, trustless raffle is *getting* the prizes to the winners. If you host a raffle in person and someone wins, you can hand them the prize and everyone will believe that they bought the winning ticket and

---

[42] Remember that a transaction id is the double-SHA-256 of the transaction's contents–all data which the person sending the transaction can modify!

should take home their prize. In an online, cryptocurrency-based raffle, you can prove that the drawing was fair and transparent, but you can't prove that the winner actually received the autographed first edition of *Modern Perl* they spent 10,000 Dogecoin to win. You also can't prove that the author's mother didn't send the winning transaction, even though she already has a copy.

To send *any* physical prize, you have to be able to associate an incoming transaction with a person, whether email address, shipping address, or something else. If you've generated a new receiving address for each person, that's easy to manage– but you have to maintain that mapping of address to people at least until you've distributed all prizes *and* you have to make all receiving addresses public so people can verify the prize distributions. This may be a lot of work. Furthermore, any mechanism that ties an identity to a transaction represents de-anonymizing data that you need to keep safe from other people. You may want to destroy this when you've finished.

Finally, any legal entity which believes itself to have jurisdiction over you may have opinions about how to run a raffle, including the idea that you can or cannot run one at all or that you or prize recipients may owe taxes on the prizes or be entitled to tax deductions based on the cost of their tickets. Your author offers no advice on this topic besides "do your research".

CHAPTER 8

More Art than Math

dogecoin and cryptocurrency build on a foundation of deep math and security to exist, run, and ensure the safety and transparency of public information. Yet you don't have to focus on the math and security and networking with everything you do. Sometimes it's more fun to focus on the odd, the bizarre, the unexpected, and the surprising.

In this chapter you'll find several ideas both fully compatible with cryptocurrency and Dogecoin *and* which bring ancillary, unique, weird, and perhaps even silly perspectives to your work with the coin.

  Tip #70   Turn Transactions into Songs

Crypto is math. At its core, we're moving big numbers around. Any meaning those numbers have is meaning that we humans put onto them. The Dogecoin network *interprets* some of those numbers. So does the Core software. Yet the value of the Dogecoin in your wallet is a consensual agreement among everyone in the Dogecoin community that a set of numbers reachable from another set of numbers and indexed by a third set of numbers represents something special and unique to *you*.

Meaning is an external property of use. What if those numbers also represented something else? What if they could create art, an image, a visualization, a game, or sound?

They can!

Generated Music

At the intersection of computer programming, music theory, and art sits something called "generated music", where a computer program assembles a bunch of numbers that match and flow sonically together in unique ways[1]. Given patterns to assemble and a starting point, the system can produce an infinite stream of audio, if you put it together correctly.

For example, Dogecoin developer langerhans[2] created a website that generates sound using data from the Dogecoin network[3].

Similarly, your author created a project with the descriptive but unexciting name of `GenMIDIPython` to do something similar with your own data.

Random Seeds and Random Meaning

How does music come from your own data?

Think about how computers think about randomness. Computers are predictable, no matter how much misbehave at the most inopportune times. Given a computer program which relies on only its inputs (it has no any hidden inputs outside of your control, like "it reads from a clock" or "it counts network packets it receives"), you should always get the same outputs for the same inputs.

When a computer generates a random number, it generally does so by using either a pseudo-random process *or* an external source of data that's sufficiently unpredictable it can serve as randomness (the time between network packets received, the way you move your mouse or hit keys on your keyboard). For more details, research pseudo-random number generation[4].

To give the *appearance* of randomness, you want unpredictability (see Embrace Entropy, pp. 21). To get unpredictability from a deterministic algorithm, you want to start from a different–and unpredictable–place each time. A *good* random number generator often has a *seed* value. With different seeds, you'll get different (apparently random) results. With the *same* seed, you'll get apparently random results but the same seed will always produce the same results.

In other words, if you provide a seed to a program and if that program initialized its random number generator with that seed, the random numbers it generates will always be the same throughout its run.

What if you used a Dogecoin address or transaction ID (a large number that's difficult to predict and essentially random enough for our purposes) as a random seed?

---

[1] Oh, yes: music is math too!

[2] https://github.com/langerhans.

[3] https://langerhans.github.io/dogelisten/.

[4] Start with Wikipedia's https://en.wikipedia.org/wiki/Pseudorandom_number_generator.

Turning Transactions and Addresses into Seeds

Are transaction and address numbers random enough? Recall the tip (see Find All Received Addresses, pp. 142) that discussed the use of `listreceivedbyaddress`. Even looking at the first 6 characters of addresses, there's a fair amount of variation in the data. While all of these Dogecoin addresses start with D, they go all over the place after that.

For more variance, pick an entire transaction ID, every bit of it. They are large numbers, and no one can predict them easily before they're created, so they provide plenty of potential uniqueness.

Making Music

Now that you have a seed (or several), install the code for GenMIDIPython[5]. When you have the code up and running, type:

```
$ python3 gen_atmospheric_chords.py <your transaction id>
['C', 'Eb', 'G'] -> C
['E', 'G', 'B'] -> E
['B', 'D#', 'F#'] -> B
['C', 'Eb', 'G', 'Bb'] -> C
['A', 'C', 'E'] -> A
['C', 'E', 'G', 'B'] -> C
['B', 'D', 'F'] -> B
['E', 'G', 'B'] -> E
['E', 'G#', 'B'] -> E
['F', 'Ab', 'C'] -> F
['G', 'Bb', 'D'] -> G
['F#', 'A', 'C'] -> F#
['B', 'D#', 'F#', 'A#'] -> B
['F#', 'A#', 'C#', 'E'] -> F#
['E#', 'G#', 'B'] -> E#
['F', 'Ab', 'C'] -> F
```

If your transaction ID is 12345abc, this will write a MIDI file named *atmospheric-chords-12345abc.id*. If your transaction ID is that exact number, you should get that exact output, at least if you're running the same version of Python and its libraries that your author used.

What can you do with this file?

The VLC media player[6] can open this file on most desktop operating systems and

---

[5]See https://github.com/chromatic/GenMIDIPython for installation instructions.

[6]https://www.videolan.org/vlc/

listen to it. If you prefer the command-line interface, the FluidSynth player[7] can do the same.

> ### Make This Sound Better
>
> MIDI files *do* include instrument information, but they *don't* include any samples that make a trumpet or jazz guitar sound like real instruments. FluidSynth supports instrument samples in a form called SoundFont. Choosing the right SoundFont can turn a MIDI file that sounds like an old '80s home arcade game into something cool. VLC can use FluidSynth sound fonts.
>
> For these compositions, the author recommends a science fiction or ambient synthesizer sound; you can find lots of SoundFont files online worth trying. Half the fun is experimenting with what turns something simple and basic into something smooth and enjoyable.

What Can You Do With This?

Supposedly a Russian proverb says "The marvel is not that the bear dances well, but that the bear dances at all." If you have musical training or composition experience, you might rightly say "This randomly-generated music could be a *lot* better."

You're absolutely right! The point of this tip is to show off the idea of art or composition or invention that uses the seed of an idea[8] to create something new and surprising and inspiring. If you have music theory or composition skills, you can modify the example code to follow a stronger tonality in a specific key, to add other instruments and generate melodies, to produce more than 16 bars of mostly-whole-note three-note chords, et cetera.

Just as a permissionless, distributed, global financial system can grow out of mutual understanding and agreement about a bunch of numbers computers swap, perhaps other interesting things can come from different interpretations of those numbers. Look for ways to turn them into data and patterns that make your ears perk up, your eyes widen, or your mind pay a little more attention to something new.

---

[7] https://www.fluidsynth.org/

[8] Pun intended!

Understand the Risks

If you listen to this audio without headphones, you run the risk of annoying the people around you. Alternately, you might amuse them. Enjoy wisely.

  # Tip #71　Control Your Jukebox

Maybe generating your own songs isn't your thing (Turn Transactions into Songs, pp. 217). Maybe you're happier with good songs performed by great artists.

Suppose you're throwing a community barn party, and you've loaded up a little Linux box in the corner with some bangers, including:

- "Violent Blue" by Chagall Guevara

- "Más Buena" by Alejandra Guzmán and Gloria Trevi

- "Starlight" by Muse

- "Hurt" by Trent Reznor and Nine Inch Nails, performed by Johnny Cash

- "Deceleration (Midpoint)" by The Incomple

...and a bunch of other great tunes. You're streaming video to online invitees, and you're even ready other folks to pick the right playlist order.

Your friends from out of town can't exactly put a quarter in the jukebox, but you don't have a coin slot on your media box anyway. Why not feed two birds with one scone and show off your Dogecoin transaction skills and let people pick songs with Dogecoin?

Set Up Your Song List

Other tips show how to turn incoming transactions into events, but this example is a little bit different. Rather than paying for a physical good (a candy bar) or an immediate single-person service (a single credit on a pinball machine), a jukebox can entertain or annoy a bunch of people at the same time (half of the world loves "Sweet Caroline" and half of the world rolls their through the entire singalong) *and* a jukebox queues up songs.

Start by picking your playlist. Each song on the list needs a unique number. Start with zero and count up.

Generate an address for each song on the playlist. Store this association somewhere (see Associate Addresses to Machines, pp. 301). In a database you might use the schema:

```
CREATE TABLE jukebox_song_list (
  dogecoin_address CHARACTER(128) NOT NULL,
  song_name TEXT NOT NULL,
  song_writer TEXT NOT NULL,
  performer TEXT NOT NULL,
  song_list_index INTEGER NOT NULL
);
```

You can add, remove, or modify this schema as much as you like. These fields allow you to model the data set of songs shown earlier. Each song has an address, a writing credit, and a performing credit[9]

Populate this table. Then, set up your Dogecoin webhooks (see Post to Discord, pp. 64 for one example) so that all transactions to the associated addresses invoke your custom jukebox code. Finally, publish the list of song data and addresses so your friends online can start queueing up songs in your jukebox!

Make a Nice UI

How will people browse songs? You could print out a short text list, make a nice web interface, or create a small app–whatever you like. The interesting plumbing is in the Dogecoin and jukebox pieces. Everything else is up to you to customize to your liking!

With all of that connected, all you need is a webhook receiver to queue up songs. That's easy to write with a little bit of Python.

Build a Jukebox

What's a jukebox? It's a list of songs and a way to play them in arbitrary order. This takes about 50 lines of Python with the right modules installed (`flask`, `python-vlc`).

The Flask portion is a simple web endpoint that takes a JSON payload containing the index of a song in the playlist and queues it. The VLC portion uses Python bindings to the VLC media player[10] to queue a list of songs or media to play.

---

[9]Trent wrote it, but the Man in Black owned it!

[10]Currently at https://www.videolan.org/vlc, but verify before installing!

A Queue Design

Only one thing makes this complicated: these two tasks have to run simultaneously. The jukebox shouldn't pause the song when someone requests a new song. Similarly, the jukebox should graciously add new songs to the queue. If you're an experienced programmer, you probably recognize that this means the system has to use some kind of parallelization, whether through threads (spoiler: it'll be threads), asynchronous programming, or inter-process management.

If you're not an experienced programmer, you need to understand that a program naturally wants to do only one thing at a time, so it's important to recognize the types of problems where this is not appropriate and design them carefully so you don't introduce subtle and difficult to debug errors. If you've ever wondered why rescanning the blockchain (see Watch Wallet Addresses, pp. 157) stops you from doing other things, that's because changing the chain while you analyze it could introduce odd inconsistencies.

This jukebox has to have two independent streams of work: playing the songs in the queue in order and waiting for input from a Dogecoin webhook. These streams have to communicate, but in only one way. When the webhook activates, that stream of work needs to put a song on the queue. That's it.

Fortunately, Python's Queue data structure does this in a thread-safe way.

Inside the Code

The first section of the code loads libraries:

```
from flask import Flask, request
from queue import Queue
from threading import Thread

import time
import vlc
```

Remember to install flask and python-vlc on your own. The second section of the code initializes the Queue object and sets up the playlist:

```
work_queue = Queue()

list = [
    'Chagall-Guevara/Violent-Blue',
    'Alejandra-Guzman-Gloria-Trevi/Mas-Buena.ogg',
    'Muse/Starlight.ogg',
    'Johnny-Cash/Hurt.mp3',
    'The-Incomple/Deceleration.ogg',
]
```

Hard-coding this list here isn't wonderful; it'd be better to connect to the database shown earlier, but for the sake of this example, it's easy to understand. The next section of code defines two functions to handle playing songs:

```
def playSongs(queue):
    while True:
        if not queue.empty():
            songIndex = queue.get()
            playSong(list[songIndex])

        time.sleep(1)

def playSong(filename):
    p = vlc.MediaPlayer(filename)
    print(f'Playing {filename}')
    p.play()
    time.sleep(1)
    while p.is_playing():
        time.sleep(1)

    p.release()
```

Python will run playSongs() in a separate thread. Remember that for a moment. This function loops until the program exits. It checks whether there are any songs in the queue. If so, it uses get() to pull the index of the next song off of the queue, then calls playSong() to play the song. This function checks the queue every second.

You can modify this to play a random song if the queue is currently empty. Otherwise there will be silence until the next queued song. For a barn party, you might prefer lots of music. For a nice quiet ice cream shop, you might prefer a jukebox that only plays songs once in a while.

Not Just Marty Robbins

A bar named Rose's Cantina just outside of El Paso, Texas has a jukebox that plays the Marty Robbins song "El Paso" on repeat. Your author can personally attest to this. Add more than one song to your jukebox.

The code in playSong() is more interesting. Python's VLC bindings give you many options, but the simplest code your author could make work *and* wanted to explain creates a new player, given the location of a media file on disk (that's the filename parameter). The function then plays the file, sleeps for a second, and

224

then uses a sleep loop to check if the file is still playing (that's `p.is_playing()`). When the file finishes playing, the code releases the media player object (p) and returns control to `playSongs()`, which will use a similar sleep loop to wait until the queue has another song to play.

That first `time.sleep(1)` is necessary because VLC takes a moment to start. If you leave out that call, `p.is_playing()` may return a false value and the song won't actually play!

The next section of code sets up the webhook destination:

```
app = Flask(__name__)

@app.route('/queueSong', methods = ['POST'])
def queueSong():
    index = request.json.get('songIndex', 0)
    work_queue.put(int(index))
    return { 'status': f'Queued {index}' }
```

This code creates a Flask app and sets up a route at /queueSong where you can post a JSON body like `{ 'songIndex': number }`. Given an input, the function pushes the index onto the queue. `playSongs()` will pull that index out of the queue as soon as the currently-playing song has ended.

Be sure to add error checking to make sure that the user has provided an actual number and that the index is valid for the size of the song list (integers zero to 4 are valid for the hard-coded list).

The final section of code starts everything:

```
def main():
    queue = (work_queue,)
    worker = Thread(target=playSongs, args=queue, daemon=True)
    worker.start()
    app.run(host='0.0.0.0', port='9999')

if __name__ == '__main__':
    main()
```

This `main()` function creates and starts a new Python thread where the VLC song player code can run. Then it launches the Flask app, listening on all network interfaces on port 9999. Change this to your preferences.

What Can You Do With This?

Nothing limits you to MP3 or Ogg Vorbis files. You can play any media files that VLC supports; VLC has plenty of plugins for many file types. This code could play music videos (change what's in `list` in the example, or in the database). If you have video output set up, you can actually see the videos. Or you could add a visualization to show album art, lyrics, fireworks, or whatever else you like for music-only media.

Be cautious that a little computer on your network listening on a port can take data from anyone who can send data over your network. Consider adding authentication and/or network protection if you want to keep someone from queueing the same song over and over.

  Tip #72   Hide Transaction Data in Media

Nothing is ever all good or all bad. The public nature of every blockchains is a virtue when all transactions are verifiable by anyone, but the vice is that all transactions are public. Your privacy is only as good as your ability to keep your identity separate from your transactions.

Posting an address on your GitHub page, your Twitter profile, or in your DNS entries (see Add a Wallet Address to a DNS Record, pp. 51) reduces your privacy a little bit. That trade-off between privacy and publicity may be worthwhile—no one wants to tip the wrong developer, after all—but every bit of privacy you lose is gone forever.

What if there were a way to add friction to the process, so that regular bystanders have to do a lot more work to find your address, but the people who really need it can still get it?

Data Hiding

The Greek word *steganography* translates to "hidden writing"[11]. In practice, it means storing data secretly in other files in ways that casual observers do not notice.

Images are common hosts, because it's easy to post images on the Internet. Furthermore, most images are large (so they can contain a lot of data) and already use compression algorithms (so pixel-perfect fidelity is not a requirement), *and* people are used to seeing compression artifacts in images the same way people expect to see brushstrokes in impasto.

Steganography fits well with media files because of the way image, video, and audio compression work: the human brain fills in the gaps such that removing

---

[11]But sadly ignores the existence of ancient reptilian herbivores.

data does not harm perceived fidelity. In the case of hidden data, adding data also does not perceptively change fidelity.

You don't have to understand all of that fully as long as you use good tools.

Store Wallet Address or Other Data in an Image

First, start with an image, such as A Goofy Dog Image, pp. 227. Ensure that the image is big and bold enough to contain the data you want to store. If you're encoding a wallet address or a passphrase, a normal sized camera phone photo is probably fine. If you're encoding the entirety of Herman Melville's *War and Peace*, you might do better with the song "Close to the Edge".

Figure 8.1: A Goofy Dog Image

While it's possible to write this code yourself, it's smarter to use code someone else has written. For example, Stefan Hetzl's Steghide[12] is a good utility.

`steghide` is a multi-platform command-line utility. Use a Linux package manager or download Mac OS X or Windows binaries. To use `steghide`, you need a cover file (the unmodified image or media file), a file containing your secret data, and an optional passphrase. Suppose you're using the image in A Goofy Dog Image, pp. 227 as your cover file. Create a file called *secret.txt* with a secret inside it: perhaps the address of a wallet you want to share with other people semi-secretly.

Run this command, split into multiple lines for book formatting purposes:

```
$ steghide embed \
  -cf steganography-pug.jpg \
  -sf steganography-pug-hidden.jpg \
  -ef secret.txt
Enter passphrase:
Re-Enter passphrase:
embedding "secret.txt" in "steganography-pug.jpg"... done
```

The `-cf` option means "cover file". This is the unmodified media input file. The `-sf` option means the stegofile, or the output file which will contain the hidden secret message. The `-ef` option means the "embed file", which is the file containing the secret message.

Run this command and `steghide` will prompt you to enter the same passphrase twice. You will need this passphrase to extract the secret file.

The output file looks almost identical to the input file, as shown in A Goofy Dog Image with Hidden Data, pp. 229. If you examine the file in this book's GitHub repository[13], you'll see that the files are in fact different. Download the stegofile and run `steghide extract -sf steganography-pug-hidden.jpg` to extract the secret file. There is no passphrase; hit Enter.

You'll get a secret message. Open the file *secret.txt* with any text editor for a personal message from the author.

You can store any kind of data here. For Dogecoin purposes, this is a good way to store a wallet address. It's an okay way to store a passphrase. It's a bad way to rely *only* on hidden data; the strength of your passphase in the steganographic file is essential to extracting data from the file.

---

[12]Currently at https://github.com/StefanoDeVuono/steghide.

[13]See https://github.com/chromatic/dogecoin-tricks-book.

Figure 8.2: A Goofy Dog Image with Hidden Data

Cute Pets and Off-Chain NFTs

Even though it's no longer 2022 and the buzz in the crypto world is no longer about NFTs[14], throwing additional junk data into transactions that most miners, nodes, and users care nothing about and can't use is a good example of a negative externality, the equivalent of throwing your litter into your neighbor's yard.

If you can find a way to brand and transfer instances of media files between people *without* storing anything other than transaction data in the blockchain, maybe you can find a better use case for NFTs than the world has so far.

For example, a steganographic NFT minting process (*minting* here means associ-

---

[14]Which is probably good, because they were a problem in search of a problem in most cases.

ating an image with the blockchain) could use a third-party website to associate an image with a transaction, create a keypair for the specific transaction, then create a transaction which requires one half of the pair to unlock the transaction and encodes that key into the image using the other half of the pair. The site then provides the image and the encoding key to the purchaser.

To transfer the image, the purchaser has to provide the key in the transaction to unlock the transaction, which then allows the third-party website to recognize the transfer.

Warning: Your author has not tried this and prefers his dog photos unencumbered anyhow.

Understand the Risks

Anytime you trade data with someone else, especially in public, you run the risk of other people getting access to that data and doing whatever they want with it, including decoding any supposedly secret or hidden messages. If you're relying on the presence of data hidden within an image, a document, an audio file, or a video to prove the provenance of that data, anyone with sufficient time and knowledge could find a way to *remove* that data or spoof their own data into the file.

If you *really* need cryptographically secure communication channels, swapping photos of cute dogs and cats is a better proof of concept and mechanism of amusement than a way to protect yourself against a state actor, a hostile corporation, or a neighbor angry that you keep dumping empty soda bottles in their yard.

Beware that any *modification* to the image, such as cropping, color correction, a filter to add an old-timey sepia tinge to your dog, or recompression if you upload to an image hosting service has the very likely potential to destroy hidden data.

This is best used as a one-off data transfer method when you don't have another secure channel to use. It's also a good way to make the world look at more cute animals.

  Tip #73   Pledge to Do Only Good

People do what they do for many reasons. From outside their minds, we can only judge those reasons their actions (and, sometimes, the words they use to describe those actions).

Some philosophers think about ethics and morality in terms of harm and benefit. Did your choices and actions help people? Did they hurt people? Other philosophers define this in terms of larger benefit. Did your choices and actions help or hurt society at large? Still more philosophers define ethics in terms of the far future of humanity. What can you do now to help the greatest number of people, living or potentially living in the future?

This is a book about funny and friendly dog-themed cryptocurrency, so some of these highfalutin' concepts seem out of place. On the other hand, this is a book about cryptocurrency, so it's important to talk about avoiding scams and traps and *not* harming others.

Many of the other tips in this book include information about how to protect yourself and your interests. Remember also that Dogecoin is a community. It's a decentralized network based on trust, cooperation, and mutually beneficial competing interests. You don't have to trust any one person or node or group, and they don't have to trust you. You do have to verify that what someone asserts to be true is actually true.

We can go further, however.

What is Good (Ethically)

How do you *know* what is good? That's a deep question. Perhaps it's easier to ask about how you can tell if something is good.

Does it help people? Is writing a book about cryptocurrency in the putative crypto-winter bear market of 2022-2024 good? Will it teach people things? Is it accurate? Is it entertaining? Is it worth the money, even if the only thing people learn is that goats are ill-mannered creatures (Vote on Goats, pp. 179)?

Is it honest? Are you including the pros and cons and benefits and drawbacks as you explain something to other people? If you're simplifying for the purpose of explanation, did you tell them that? Do they understand there are other important details and do they know where to learn more?

Is it verifiable? Can you prove the truth of your claim? Can other people? Can other people prove this? Easily?

Is it sustainable? Does this thing you're doing work only once? Can other people do it? Can they do it by themselves, or at least without you?

Do the people benefiting pay the costs? If you do something that requires other people to pick up the tab, who are they? What benefit do they get? Do they give informed consent? Do they give informed consent *to your specific use case*?

These are all abstract questions. It may be impossible to create an ethical code of conduct that covers every possible situation you find yourself in. If and when you have trouble answering one of these questions unambiguously yes, ask yourself "What would need to be true for me to answer this question unambiguously yes?"

What is Useful (Practically)

Ethical and useful aren't always the same. It may be ethical to start an online donation site where each donated Dogecoin means you'll plant cilantro in your backyard to combat climate change and increase carbon capture, but it'd take a lot of cilantro to make any difference at all, and cilantro–also known as *hate parsley*–is disgusting.

What interesting characteristics of *useful* might apply?

Create more value than you capture. Writing a book is fun[15]. Writing a book and having other people *read* it and learn things and do cool things is, hopefully, an example of a value-producing activity.

Share knowledge and capabilities. This is one of the most valuable characteristics of free and open source software. A seasoned software developer might find a very interesting new feature or effective optimization and keep it to themselves, but submitting a pull request to one of the numerous Dogecoin projects puts this new capability in the hands of other people who can also benefit from it.

Make it easy for other people to do the right thing. Again, *right* is difficult to define, but ask yourself if you were trying to do something and you weren't an expert, how likely would you be to get the outcome you want without running down blind alleys or accidentally making costly mistakes?

Good In Practice

What does this look like in a specific context? Let's talk about the responsibility of a core maintainer of important Dogecoin software, such as the Dogecoin Core.

At the minimum, can we agree that core maintainers should:

- Review all code changes carefully for correctness, security, and maintainability

- Review all bug reports, feature requests, and issues to understand proposed problems, actual problems, and workable solutions

- Disclose any potential or perceived conflicts of interest to other maintainers as they might be material

- Strenuously avoid discussions of price, investing, and speculation that could be seen as attempts to manipulate the market

---

[15]For some people.

232

The last is tricky. It's hopefully ethical to discuss potential market cap (Calculate Dogecoin Market Stats, pp. 245) and to acknowledge that Dogecoin can be used to make money (Make Money with Dogecoin, pp. 252), provided that one discloses any biases and keeps the discussion focused on verifiable facts and figures.

It's hopefully also obvious that it would be unethical to know of potentially exploitable security flaws or network limitations and not disclose them to other maintainers or, worse, actively exploit them for personal gain.

In other words, a position of trust and responsibility in this context–or any context– includes the requirement to deal honestly, avoid hiding information, and to prioritize the health of the network and the community over personal gain.

Those seem like good guidelines even apart from the context of maintaining software or Dogecoin itself.

Understand the Risks

Doing the right thing isn't always easy. It's tempting to take a shortcut. It's disheartening to see careless, sloppy, or unethical people brag about how quickly they were able to get ahead even if you know they stepped on other people to get there.

You have your own definition of success and rules for living. Your author hopes to live his own life in such a way that he can wake up every day, look at the world, and see that it's a little better for his participation in it. He hopes to go to bed every night, reflect on the day or week or year and see that he's treated other people fairly and honestly, with kindness.

That doesn't always happen. People make mistakes. Even with the best intentions, things go wrong. Perhaps the last and final test of ethics is if you can own up to your mistakes–even unintentional–and do what you can to set things right.

Keep on doing only good everyday, and may your every day end with the sleep of the just.

  Tip #74   Hide Addresses in Stories

Steganography in images can be fun (see Hide Transaction Data in Media, pp. 226), but it's not the only way to obscure something in plain sight. As the world of machine-generated content improves, you can use something called *prompt engineering* to build something new or new-ish out of an existing set of data.

Let's put together a couple of unique ideas to make something that didn't exist before: an AI-powered short story generator based around a representation of a specific Dogecoin address.

Or, put another way, let's ask "What if Jules Verne lived at my (Dogecoin) address?"

Encoding an Address with BIP-39

The BIP-39 wordlist, often used to represent passphrases for private keys, is a list of 2048 words intended to be easy to use, read, remember, and distinguish from each other. Starting from a random number (see Embrace Entropy, pp. 21), you can turn produce a BIP-39 compatible list of words–and vice versa. While the original intent was to give you a list of words that's easier to remember than a long random number[16], it's also a list of valid words in many languages, including English.

Start with a source of random data, such as a sufficiently long random number. How about a Dogecoin address? An address can become 21 bytes of data with Base58Check decoding. Given 16, 20, 24, 28, or 32 bytes of random data, you can generate a BIP-39 mnemonic. Given 21 bytes, that means we need to remove one and hold it for later. Hold that thought.

More Entropy

More entropy is usually better, so using Base58 encoding *without* the 4-byte check data would give you more entropy, but you still have to remove one digit to make the number of bytes work with the mnemonic encoder. Also be sure to decode and encode with the same mechanism so you don't lose information.

A bare-bones Python program to perform this encoding could look like:

```
from base58 import b58decode_check
from mnemonic import Mnemonic

address = "DAY5wNkebzEyqUXCkN9koKNBuzXRKRTjcL"
mnemonic = Mnemonic("english")
address_bytes = b58decode_check(address)
words = mnemonic.to_mnemonic(address_bytes[0:-1])

print(address)
print(words)
```

---

[16]See https://xkcd.com/936/, for example.

234

This code uses the `base58` and `mnemonic` modules from PyPI to decode the hard-coded address into a byte array, then convert that into a BIP-39 mnemonic. The interesting code is the expression that uses all but the last byte of `address_bytes` as input to the `to_mnemonic` method. With this approach, there are 20 bytes of entropy.

Be aware that the mnemonic generation is deterministic; given the same input, you'll always get the same output. This is really useful.

With this list of words, now you can do something interesting with it.

Text to Text with LLMs

Services claiming to be artificial intelligence have been around for a long time, but in practice they're systems which use statistics to turn a set of inputs into a set of outputs[17]. As of 2024, you can use any of several services such as Claude, ChatGPT, Gemini, and more to generate text from a prompt.

These services are interesting because you can give them a short prompt and they'll produce output you can refine, edit, improve, or otherwise use as the start of something more detailed or useful or interesting. Newer LLMs and machine learning systems have ever-more training data, so they have a better chance of producing something close to what you want than older versions.

After you've turned your address into a wordlist, try a prompt such as "Write a short story in the style of (H.P. Lovecraft, Flannery O'Connor, Jules Verne, Lois McMaster Bujold) including each of these words in order: abandon ability able…".

You'll get back something that reads like:

> The bullet, swift as sunset, pierced the air over the hockey rink. A fatal shot, no chance for escape. The brown ice bore witness as the captain fell, a chapter of glory ending abruptly.
>
> In the midst of this frozen battle, a cricket's chirp rose from an alley nearby.

A bright sound against the cold clash of forces on the rink. Water droplets clung to metal railings, a reminder of the mystery of life and death, of victories and losses.

> An exchange of glances among the players, a wordless understanding passing like a secret note. The alley's mystery mingled with the aggression, making the violence on the ice seem strangely distant.

---

[17] Your author prefers the term "machine learning".

That's not *exactly* Ernest Hemingway, but it's also not obviously William Faulkner either, so it's a start anyhow.

Now what do you do with it?

Text to Address

To decode this, someone needs to extract all of the relevant words in the proper order (that's why the chat prompt asked to use them in order), then add on the missing digit, then run the data through the BIP-39 decoder. Whew.

```
from base58 import b58decode_check
from mnemonic import Mnemonic

mnemonic = Mnemonic("english")
new_address = mnemonic.to_entropy(words)
new_address.append(missing_byte)

print(bytes.decode(b58encode_check(new_address)))
```

This code is the opposite of the previous code. Given an array of words in `words`, the `mnemonic` object converts the words into a byte array representing the original entropy, the 20 bytes of data making up almost all of the Dogecoin address. Append the missing final byte, then encode the result as a Base58Check string. Finally, turn that Python byte array into a string and print it and there you have the original address (or whatever you encoded).

To make this work in practice, you *do* need a way of getting that last byte to the recipient, otherwise they'll have to loop around all of the 256 possibilities to find it and may get strange results. That means this isn't *effective* data security, because it can be so easily brute-forced, but it may be acceptable for your purposes.

Understand the Risks

As with any form of steganography, any data you release publicly is out of your hands. Relying solely on people not knowing to look for hidden messages is very little safety against curious codebreakers. Also be careful about what information you share. Unless you're running your own instance of an LLM-powered text generator, you run the risk of the server operators seeing and logging your prompt. Even if that's only a public passphrase, you've exposed information.

Hiding your *private* key–especially the mnemonic–in this form is a very risky idea. Hiding a key for a scavenger hunt or puzzle (Host a Treasure Hunt, pp. 174) is better; you expect people to decode your message eventually, with the appropriate amount of work.

If you encode a *receiving* address, you risk some of your identity and privacy but not your funds. Anyone who recognizes the words from the BIP-39 wordlist on sight[18] will be able to reverse-engineer *something*.

If, however, you use this mechanism to write up something nice inside a birthday or graduation or congratulations or feel better or just because card, you're probably fine. After all, Flannery O'Connor won't be writing a new short story any time soon, but you can do something nice for a friend.

Speaking of specific authors, be aware that many public generative AI services have been credibly accused of training their models on copyrighted data without the consent of the copyright holder. Work that's obviously in the public domain (Jules Verne, Mary Shelley) is up for grabs, but the work of a current, living, writing author (this book's author, for example) may have been used with neither consent nor notification (this is in fact true of this book's author).

Choose your source material carefully.

Tip #75   Create Crossword Rewards

If you like to start or end your day with a little bit of cruciverbalism[19], there's a way to combine your interest in puzzles with the blockchain.

What's in a crossword puzzle? You need a few things:

- a list of words and clues

- a grid

- a way to fit words into a grid

Fortunately, the Internet provides. Saul Pwanson's wordlist[20] is one of many good sources. The *xd-clues.zip* file in particular contains plenty of words and clues, going back decades.

---

[18] Admittedly, that's a small number of people who have them memorized, but that group of people is also *motivated*, so the less you raise their attention, the better.

[19] Crossword puzzles!

[20] See https://xd.saul.pw/data/.

Another good project is genxword by David Whitlock and Bryan Helmig[21]. Given a list of words, this program can produce a PDF, SVG, or PNG file with a crossword puzzle, including a list of clues and the word bank used in the puzzle.

It's easy to tie these things together.

Generate a Crossword Puzzle

Given the wordlist and genxword, you can generate a crossword puzzle. First, grab a random sample of around 100 words and clues. Of course, this is complicated by the fact that the wordlist is a tab-separated text file, with the first two columns the source (New York Times, Wall Street Journal, etc) and the year of use.

First things first, get 100 random lines from the file with the Unix shuf utility:

```
$ shuf -n 100 clues.tsv > daily-clues.txt
```

Now translate the tab-separated file into a list of words and clues by removing the first two columns and using a space to separate the words from the clues[22]:

```
$ cut daily-clues.txt -f 3,4 --output-delimiter ' ' \
  > daily_word_list.txt
```

Finally, run genxword on the word list to generate a letter-sized PDF file:

```
$ genxword daily_word_list.txt -o daily_puzzle l
```

See the genxword documentation for more options. With this command, you'll get two files, *daily_puzzle_gridl.pdf* and *daily_puzzle_keyl.pdf*. If you want a textual list of clues and words, use the n option to produce PNG files instead or in addition.

Now you have something you can share with other people.

Rewards on the Blockchain

What happens next? Add a reward! Assume you've put 100 Dogecoin aside for this. Hold the question of how to fund that prize for a second. First, think about how someone will claim that reward. Obviously they need to start by solving the puzzle. Then you have at least two options.

---

[21] See https://github.com/riverrun/genxword.

[22] The puzzle generator prefers this format.

Solution as Private Key Input

If you use some or all of the solution as the input for a private key, a solver can add the private key to their wallet and claim the reward by transferring the funds to an address under their control.

If you use this approach, make sure the rules for extracting the key input from the solution is clear and unambiguous. For example, you could use the rule "Join all across words together with spaces, then join all down words together with spaces, all lowercase". Or you could use the rule "Take the first letter of all across words, then the first letter of all down words, all uppercase". Whatever approach you take, make it unambiguous and easy to verify.

Then document the approach you yourself used to generate the private key and address from the seed phrase. Is it a wallet generator? Is it libdogecoin? Is it something else? (Perhaps Make Keys from Answers, pp. 240?)

One of the last things you want to do when putting together a puzzle is to mediate a dispute between multiple people who all claim to deserve the reward but can't agree on the solution.

Solution as Transaction Unlock Script

An alternate solution–untethered from keys and addresses–is to use something derived from the puzzle solution as the unlock script for a transaction. As with the private key mnemonic in the previous example, the goal is to make something no one can guess without solving the puzzle.

Your rules can be similar: "Join all across words together with spaces, then hash this to generate a value which is the unlock script". For this to work, you need to create a P2SH transaction (see Decode a Transaction Script, pp. 203). Use SHA-256 to hash the solution, then use SHA-256 on it again. The result is the unlock script, which will look something like this:

```
OP_HASH256
... (your hashed value)
OP_EQUAL
```

To unlock this transaction, the solver must provide the solution to the puzzle (hashed once). Yes, you read that right: you've hashed the solution three times, once to turn a crossword solution into something you can use to unlock a transaction, and then twice more to lock the transaction.

There is one downside to this approach: if someone solves the puzzle, anyone can look at the pending transaction in the mempool and submit a similar transaction and then it's a race to see which pending transaction gets mined into a block first. There's no mitigation for this on the network itself, so you might have to put rules around it, such as "submit your answer to an arbitrator first in the case of front-running" or "don't let jerks ruin the fun for everyone", but both of these are difficult to enforce and may reduce user privacy.

Either approach you use–private key or unlock script–you'll need to fund the prize by transferring funds to the appropriate address. Once the transaction has settled, publish your puzzle to the appropriate people and let them have at it!

Understand the Risks

If there's money involved, there's always the risk of someone cheating, or at least bending the rules. If you use a public wordlist with a known set of words and clues, it's possible for someone to write a program to solve any arbitrary puzzle by mapping clues to words. Mitigating this risk might mean making your own clues, adding words from a private list, or otherwise making it more difficult to go from a solved puzzle to the solution to unlock the reward transaction.

Similarly, if you're using random numbers to generate the puzzle, it's possible for someone to reverse-engineer the random number generator and predict the clues used. Every time you see the word "random number generator", do your diligence to ensure that the numbers are random and have enough entropy (see Embrace Entropy, pp. 21) to guard against prediction attacks.

If you're publishing puzzles publicly, remember that the clues and answers are often language- and culture-specific. Referring to a television advertisement for chocolate milk in the 1980s featuring Pancho Pantera[23] might exclude a broad swath of your potential audience. Furthermore, if you reliably publish your puzzles at 10 am local time, people who are awake and not at work right them might have an advantage solving puzzles over people who have to wait several hours to have time free to ponder your clues.

Most of these risks summarize to "help everyone compete fairly".

  Tip #76 Make Keys from Answers

Suppose you've gone to all the work of making a year's worth of weekly crosswords (see Create Crossword Rewards, pp. 237) and you want to publish them and let your community meet the challenge for the chance at rewards: Dogecoin transactions they can claim as their own.

How do you *actually* do that in practice?

---

[23] See https://www.youtube.com/watch?v=4oMQSGRnNiY.

A Puzzle Solution to Key Pipeline

Suppose you have a simple puzzle with four words in it, two across and two down. Starting with the Across words, reading in increasing numeric order, then moving to the Down words, also reading in increasing numeric order[24], the word list is:

- ARGYLE

- TEATIME

- BOWTIE

- SQUAMOUS

That's the puzzle solution. Now you as the puzzle maintainer (and everyone who solves the puzzle) has to turn that data into the key to unlock the puzzle's rewards. You're already halfway there, though: you have a list of data that's ideally unique and hard to guess. Now you need to turn that into a key.

Data to Keys

Normally this approach is a bad idea because the security of a key is essential, but where you *do* want to share a key, this approach works. Think about BIP-39 (see Hide Addresses in Stories, pp. 233): that proposal describes a mechanism to turn a list of words into entropy used to generate a private key.

The same approach works here by using the words to *create* entropy, or at least sufficient entropy that people are unlikely to guess the private key without solving the puzzle.

Remember also that this process must be *repeatable* such that anyone who's solved the puzzle should be able to generate the same private key and have a chance to redeem the reward.

One reliable way of generating a private key is to turn the wordlist into a hashed value, then use that as the source of data for a new key. This is, more or less, what BIP-39 does, though that standard has some advantages that this approach doesn't use. In particular, BIP-39 word lists reduce ambiguity and the rest of the standard provides some convenient security features which multiple clients and multiple cryptocurrencies have adopted.

---

[24]Yes, this explanation seems tedious, but if you're hosting a competition, the clearer the rules, the fewer arguments you'll have to mediate.

While the puzzle transaction approach is good and may seem more in line with the goals of the blockchain, there's no simple way to create an unlocking transaction without creating at least part of the transaction manually. If your audience is highly technical and capable of creating and signing their own transactions, and knows to test rigorously on testnet (see Test First, Safely, pp. 193), this approach may work. For general users, the risk of leading them to tricky and potentially coin-losing mistakes is high.

Suppose you were to write a little Perl[25] utility to do this:

```
use Modern::Perl '2024';

use Bitcoin::Crypto::Key::Private;
use Digest::SHA 'sha256';

my $bytes = sha256( $input );
my $key   =
    Bitcoin::Crypto::Key::Private->from_serialized( $bytes );

$key->set_network( 'dogecoin' );

say $key->to_wif;
```

Given the Perl libraries `Modern::Perl`[26], `Bitcoin::Crypto::Key::Private`[27], and `Digest::SHA`[28], this program takes a single argument, hashes it, and turns the hash into a private key. Then it prints the private key's WIF (see Interchange Your Wallet Keys, pp. 15).

Run this with a single argument, in this case a quoted string[29] of the four words in the puzzle solution:

---

[25] Use any language you trust and your users are likely to use. In this case, your author has used and contributed code to this library before, so he trusts it.

[26] Enable convenient features.

[27] Manage keys for cryptocurrency networks.

[28] Use cryptographic hashes.

[29] If you don't quote the string, the shell will treat each word as a separate argument.

```
$ perl wif_from_text.pl "ARGYLE TEATIME BOWTIE SQUAMOUS"
QPBcDcDxMnXcuts7c3VcXPU35BQxRn9iAoUSRETXDMligu9MEDkW
```

Now anyone can import that WIF into any compatible wallet.

What Can You Do With This?

As written, the puzzle's reward is available to anyone who can solve the puzzle, generate the private key, and create a transaction to move the unspent funds to an address of their preference. As soon as you publish the puzzle and the key generation rules, the race is on.

To give everyone a better chance of competing, you *could* fund the reward with a timelocked transaction (Timelock a Vault, pp. 200), so that the race is between transactions, not people. Anyone who has the solution could import the private key and attempt to spend the funds, but the timelock will mean that no transaction can move the funds until the time expires. At that point, if one or ten or a hundred competing transactions all are in the mempool, whichever transaction gets mined into a block first will win.

Think of this as a random drawing for all winners.

Regardless of *which* technique you use to provide rewards on the blockchain, having funds sent to a puzzle address or to an address for which people allows you and/or anyone else to increase the puzzle's reward. Publish the address and anyone else can send funds there as well.

If your users aren't likely to have Perl, Ruby, Python, PHP, Go, Rust, Node, or whatever language you coded your solution, consider creating a web app where they can paste their words. Or you could go all the way and create a full-fledged web app that allows them to solve the puzzle online and generate a WIF for them.

Of course, remember that this technique for generating keys is not secure because the *problem* you're trying to solve here is not primarily a security problem. You *want* other people to get access to the key and its funds. Never use this technique for your own private funds you want to keep secure; it's better to stick with a hardware wallet, a BIP-39 generator using sufficient entropy, or another technique not intended to be guessed or shared.

If you do use this technique, make sure you test it thoroughly. Also take care to tell your users that they should never rely on a key generated from software they can't inspect running on computers they don't control and trust for their own funds.

## CHAPTER 9

Dogecoin in the World

Watching numbers change on a computer can be soothing or fulfilling for a while, but eventually the physical needs of being human intervene. While Dogecoin's originally poked fun at cryptocurrency and people who take cryptocurrency far too seriously, the very real human needs of utility (including having fun) are serious business.

This chapter explores what Dogecoin can mean in the world. A word of warning, however: some people treat cryptocurrency and Dogecoin as a get-rich-quick investment. This book takes a different approach. Good things are *useful* and *enjoyable* for reasons beyond numbers going up.

  Tip #77   Calculate Dogecoin Market Stats

In 2013, your author received his first Dogecoin tip, probably for saying something insightful or clever on a social media forum. It was worth a fraction of a penny then but it meant something–it gave your author a warm glow that he'd brought something meaningful into the world.

In 2021, that tip was worth hundreds of dollars. At the time of this writing, it's worth tens of dollars. Who knows where it will be in a week, a year, or ten years?

Some people care a lot about that. If you're using Dogecoin to buy a meal or donate to pay for medical costs of a young child with illness or to support a friend on the opposite side of the world, the fluctuations over time aren't important. If you also or otherwise see an asset that may appreciate over time, you owe it to yourself to do the math to figure out what *could* and *won't* happen, if fortune favored you.

Count All the Doge in Existence

How many Dogecoin are in the world? To answer that question, use any reputable block explorer (or ask a search engine to find one). As of this writing in June 2024, multiple sources suggest there are about 144 billion Dogecoin in the world. This number grows by ten thousand with every block minted, which happens about once per minute. On average, expect miners to find about 5.26 billion new Dogecoin every year.

As of the current writing, there are about 8.12 billion people in the world.

This means that there are, at most, about 17.7 Dogecoin available for every person in the world. Maybe you have a few more. Most people have fewer. If the population stayed flat at 8.12 billion, everyone could get 1.5 more Dogecoin next year.

That's not a lot of coins to swap between people if you want to buy cool art or delicious beverages, to pay someone to write you a fun computer program, or to reward someone for answering a difficult question. Let's think of Dogecoin in a different way.

Count All the Value in Existence

If you sold 100 Dogecoin at the highest possible valuation Dogecoin has reached (as of this writing), you'd have about $70 USD. If you sold 1 Bitcoin at its highest possible valuation as of this writing, you'd have about $73,800 USD. That's a big difference!

Only 21 million (million with an m, not billion with a b) Bitcoin will ever exist. Right now, if all of the Bitcoin that could ever exist existed, there'd be 6,857 Dogecoin for every 1 Bitcoin. If Dogecoin had the same market capitalization as Bitcoin at its all time high, that $73,800 per Bitcoin would be equivalent to about $10.76 per Dogecoin. The math goes like this: 144 billion Dogecoin divided by 21 million Bitcoin gives 6,857 Dogecoin per Bitcoin. $73,800 divided by 6,857 gives $10.76.

What if there were no Bitcoin? What if all other cryptocurrencies and currencies in the world vanished, leaving only Dogecoin?

The World Bank estimates the total Gross World Product–the value in dollars of all of the economic transactions of all of the countries of the world–at $96.5 trillion (that's trillion with a t) US Dollars at the end of 2021. Let's subtract two years of Dogecoin from the current number of coins to give 133.74 billion.

If *every one of those dollars worth of GWP turned into value in Dogecoin*, each Dogecoin would be worth $721.55 USD.

Understand the Risks

If you investing in Dogecoin and hope it reaches $10,000 per coin so you can buy a nice island and shoot rockets at Mars, you'll have to wait for the GWP to grow.

These calculations are rough. They don't take into account things lost Dogecoin–sent to addresses with forgotten keys or addresses that don't exist. They don't account for changes in mining strength where there may be more or fewer blocks mined than expected.

They *do* help to set expectations. The entire world economy may never switch to Dogecoin as a world currency in the lifetime of anyone reading this book. All 8 or 10 or 20 billion people in the world may never get their average of 17.5 Dogecoin per person. Your wallet of 100 Dogecoin may never be worth $7,380,000 and you may not be able to sell it and buy the goat farm you've always dreamed about nestled between Alberta's wheat fields and mountains.

Dogecoin can still be fun and useful. If anything, realizing that a billion-dollar moonshot is mathematically and economically impossible frees this goofy cryptocurrency to do interesting and quirky and amusing things. It doesn't have to be the buttoned-up, why-so-serious currency of Very Serious People who scowl as they mark up charts and talk up their bags.

It's free to be the currency of people who enjoy creative fun while doing useful things.

Be creative. Have fun. Do useful things.

  # Tip #78   Track Blockchain Sizes

You never have to ask permission to participate in Dogecoin, but your ability to participate may have a cost. To run a node or a miner, you must account for electricity, bandwidth, disk space, and routine maintenance to keep things running. Sometimes this is a thankless task, but it helps you and the community.

What does it take to make the network work? We can measure that!

Count All the Unspent Transactions

Other tips explain how important the entire blockchain is essential to proving that any given transaction belongs to the network (see Forge a Chain, pp. 7, for example). Even though every block in the chain validates every future block, some blocks and transactions supersede others. Any transaction where the recipients have spent all the outputs (see Bundle and Track Transactions, pp. 163) is useful for verifying historical accuracy, but it's not essential for checking the validity of any new proposed transaction.

In other words, to validate a new transaction, the network has to prove that its inputs (the coins it proposes to spend) all come from unspent outputs of existing transactions. These are *UTXOs*, or Unspent Transaction Outputs. A node doesn't have to track every transaction or block everywhere to validate any new transaction; it only has to track all UTXOs.

The set of all UTXOs grows and shrinks over time. If you receive 10,000 Doge from mining a block and then create two transactions, one sending 5,000 Doge to your cold wallet and another sending 5,000 to your favorite Uncle Oscar, you've replaced one transaction with unspent outputs (the coinbase transaction) with two (one to your cold wallet and another to Oscar).

If you've received a hundred tips to a wallet address and you decide to sweep them all to a cold wallet in one transaction, you may replace 100 UTXOs with one[1].

There's no *theoretical* limit to the maximum total number of UTXOs. Instead there's a *practical* limit of how many the average node can store. To see what's currently happening in the network, use the `gettxoutsetinfo` RPC command to inspect the *current* set of UTXOs. This will take several seconds to return a result resembling:

```
{
    "height": 4667058,
    "bestblock": "5fe7bc...",
    "transactions": 8281767,
    "txouts": 17644750,
    "bytes_serialized": 769053979,
    "hash_serialized": "eae241...",
    "total_amount": 138843085485.95428675
}
```

This output shows 8.28 million transactions with unspent outputs and 17.64 million unspent outputs in those transactions, with 138.843 billion (billion with a b) Dogecoin available for spending.

Count All the Dogecoin

Want to know how many Dogecoin are available? You just did! Great work!

Although it gets *slightly* more complicated: not all of these coins can be spent. People may lose their wallets and keys. Some coins may be locked up until a point in the future. Others were sent to addresses where there can be no keys. Even still, an estimate is a good start.

---

[1]Note that this example doesn't account for transaction fees paid to miners.

Measure Disk and Memory Consumption

Disk space and memory aren't free. What does it cost to run a node? For example, how much memory does your node use to track UTXOs? Use `getmemoryinfo` for a hint:

```
{
  "locked": {
    "used": 14687328,
    "free": 254880,
    "total": 14942208,
    "locked": 14942208,
    "chunks_used": 458979,
    "chunks_free": 3
  }
}
```

This node uses 14,687,328 bytes of memory of 14,942,208 bytes available. If you're running the GUI, you'll need more than that, but this is a start of an estimate.

How much disk space are you using? Again, an exact example is difficult to get. With Dogecoin Core 1.14.x, the `bytes_serialized` value returned from `gettxoutsetinfo` shows the size in bytes of the UTXO set as stored on disk, more or less. In newer versions of Dogecoin Core (1.21.x and newer), this single value will become `bogosize` and `disk_size` with more accuracy. In the previous example, the Core returned a value of 769,053,979 bytes: around 7.7 MB on disk.

If you're running a node, you can use your operating system's file utilities to get a better sense of how much space on disk your node uses to store the blockchain. As of this writing in early April 2023, a full node consumes about 64 GB (gigabytes[2]) of disk space.

Understand the Risks

Besides the risks of running a Core node that may lead you to separate your wallet from the node, you need RAM, disk space, and bandwidth to participate in the network. These aren't free. Neither is electricity. Your participation is valuable–essential, even. Without multiple cooperating well-behaved shibes, a few bad actors could control all of the transactions.

Consider how much you're willing to invest in keeping Dogecoin independent and permissionless. Use what you've learned here to monitor what's happening regularly, lest any changes catch you by surprise.

---

[2]That's gigabytes with a hard-g, not a soft-g like GIF.

Tip #79 Calculate Your Dogecoin Foot-print

Running a node isn't free, unless someone else paid for your computer and hard drive and continues to pay for your Internet connection, electricity, and occasionally brushing the dust off of your computer. Maintaining a node is sometimes a thankless task, but it keeps the network healthy.

If you hold Dogecoin, you're asking the network to keep track of your coins. A tiny bit of every node's memory and storage space keeps your money tracked appropriately. How tiny? How large? It depends on how much you use. You can figure this out!

Count Your Transactions

Your footprint–the amount of cost you generate on the network–shows up in a couple of ways. First, the total number of transactions you've transacted affects storage space. Every transaction gets recorded and sticks around for posterity, after all. Every full node keeps a copy of every transaction in every block, so the more transactions, the more storage every full node needs.

How many transactions has a wallet transacted? The `getwalletinfo` RPC command returns a field, `txcount`, showing this value:

```
{
  "walletversion": 130000,
  "balance": ...,
  "unconfirmed_balance": 0.00000000,
  "immature_balance": 0.00000000,
  "txcount": 133,
  "keypoololdest": 1618872149,
  "keypoolsize": 100,
  "paytxfee": 0.00000000,
  "hdmasterkeyid": "..."
}
```

Count Your Unspent Transactions

The second way your activity affects the blockchain is the amount of data every node has to track. Every node, even if it's not a full node, needs to know about every transaction with unspent outputs (UTXOs). This allows nodes to validate that any attempt to send Dogecoin to an address gets those Dogecoin from a valid input.

How many UTXOs are in your wallet? The `listunspent` RPC command returns information about every unspent transaction associated with the current wallet. In the console or from the command line, you'll get a list of transaction data formatted as a JSON list:

```
[
  {
    "txid": "37640d...",
    "vout": 0,
    "address": "DQniBb...",
    "scriptPubKey": "...",
    "amount": 5.00000000,
    "confirmations": 702433,
    "spendable": true,
    "solvable": true
  }
  ...
]
```

If you use `dogecoin-cli` and pipe the output to `jq` (as seen in Find All Received Addresses, pp. 142), you can easily count the number of unspent transactions associated with this wallet:

```
$ dogecoin-cli listunspent | jq '.[].amount'
13
```

In this example, the current wallet has a handful of unspent transactions. Remember, you can use any `jq` command on this data structure you want. For example, use the command `jq '.[].amount'` to extract the *amount* from every transaction.

Estimate Your Footprint

How much of the unspent transaction output list represents your unspent transactions? Use `gettxoutsetinfo` to find the number of total UTXOs (see Track Blockchain Sizes, pp. 247), then divide:

```
$ dogecoin-cli gettxoutsetinfo | jq '.txouts'
17657052
```

In this case, divide 13 by 17657052 and multiply by 100 to find out that this wallet represents 0.000073625% of all UTXOs. Put another way, the next 15 minutes of mining blocks will generate more UTXOs than this wallet has.

What's your footprint for the total number of transactions? As of this writing, a good estimate for the total number of transactions is 95 million. 133 divided by 95 million is 0.00014%. Again, this is a wallet with a tiny footprint.

251

Understand the Risks

What's a good footprint? It depends. A wallet with one transaction holding a billion Dogecoin takes about as much space on a node as a wallet with one transaction holding one Dogecoin. A wallet that's transacted a million times but only holds 100 Dogecoin in unspent outputs takes up a lot of space on full nodes but may only have a couple of UTXOs, while a wallet with 100,000 Dogecoin received in two million tiny transactions will take up a double-digit percentage of the UTXO set.

If you find yourself in any of these scenarios, should you do anything[3]? It depends! If you realize you're taking up a lot of storage space, you can consider running a full node. Obviously you're getting a lot of value from the Dogecoin network, so you can pay it forward.

If you have a large number of unspent transactions, consider spending those coins. Do something nice for yourself. Tip a friend. Consolidate your funds, even. In any case, remember that "owning" a bunch of numbers that a network of computers over the Internet shuttle back and forth buys you bragging rights until you do something productive with them. Hoarding for the sake of hoarding costs other people time and resources. Ask yourself what good you could do in the world, then do it! That's the spirit of Dogecoin.

  Tip #80   Make Money with Dogecoin

People do the things they do for many reasons. Sometimes you want to impress other people. Sometimes you want to feed yourself and your family and a bunch of animals. Sometimes you want to learn something new, or spent a couple of hours doing something different. Sometimes you see a problem your brain won't let you leave unsolved.

Even as much as the Doge community's motto of Do Only Good Everyday is a lovely ideal, other wants and needs motivate people's words and deeds. Furthermore, half of the word "cryptocurrency" is "currency". That means money.

Ignoring money ignores important motivations that move people. It's time to talk about money: specifically, how to make money with Dogecoin.

---

[3]Besides sending a significant portion of that billion Dogecoin to buy a copy of this book for everyone you know.

Sell Something Cool

This book tries to reinforce the theme that *you need no permission to do something cool* or fun or amazing or useful. You don't need to run a Core node or any software at all to receive Dogecoin as payment. All you need is an address to which you can receive transactions. You can calculate that by hand or export it from a node or use any other mechanism you prefer–as long as you keep your private keys safe and secure.

Suppose you keep bees and sell honey every spring and summer weekend at the local Farmer's Market. Why not print a QR code on your stand (see Generate a QR Code, pp. 146) and experiment with taking payments in a different currency? You will face some challenges, such as getting the price right based on the current exchange rate, waiting for the payment to settle, and explaining why a cartoon dog wears a bee costume (see Customize Your QR Code, pp. 149). Maybe it's worth it to overcome those questions.

Maybe you charge admission for something, like a pinball arcade (see Watch Wallet Addresses, pp. 157), where proof of entry is proof that a customer-specific address customer has received a payment. This avoids some questions (When has my transaction settled? It's a pre-event purchase! Buy your ticket at least a couple of minutes in advance!) and raises others (What if the cost of admission fluctuated dramatically between purchase and event?).

To make a million Dogecoin by selling:

- pinball credits at 10 Dogecoin each, sell one hundred thousand plays

- jars of honey at 100 Dogecoin each, sell ten thousand jars

- concert admission at 250 Dogecoin each, sell four thousand tickets

Is a million your goal? Maybe. Maybe not. Do the math for your own situation.

Become an Expert

Perhaps you're less an apiarist than you are a music tutor, editor, or sketch artist. Perhaps you have certain programming skills from years of deciphering open source code. Your time and experience have value, and you deserve compensation for that value at a fair and equitable rate.

Taking payments from outside of the country can be tricky. Taking payments from individuals can be slowly and risky. A quick, permissionless, decentralized system for sending payments over the Internet can remove friction from this process[4].

Maybe you've set up a face-painting or caricature stand at a school carnival, or your kids have a "muddle your own blueberry lemonade" stand on the street corner on a hot weekend day, or there's a line under the sign you jokingly hung over your back porch reading "5 Minutes of Advice, Line Starts Here" and you realize your time is worth something.

To make a million Dogecoin by:

- selling the experience of making blueberry lemonade at 25 Dogecoin each, help people mix forty thousand glasses

- selling open source consulting at 1000 Dogecoin per hour, consult for one thousand hours

- painting Cheems on cheeks for 40 Dogecoin each, send forty thousand happy dogs into the world

- mowing lawns for 35 Dogecoin each, mow twenty eight thousand small lawns

Invest in Usage, not Value

Many of these examples and countless others are within your reach right now. If you brainstormed for an entire weekend, you could probably find at least three or four Dogecoin-generating ideas with what you already know and have available.

If your goal is to turn your efforts into cash in the bank, you need to know two things. First, what's the currency value of your current unspent transactions. Second, what's going to happen in the future?

The latter question is easier to answer: no one knows. If the maximum potential exchange rate for Dogecoin is about $720 (see Calculate Dogecoin Market Stats, pp. 245), you need to hold 1400 Dogecoin to have the value of a million dollars. For *that* to happen, a single Doge has to be worth 5000 times what it's worth right now, as of this writing.

Is that likely to happen? Don't take anyone's word for it. Anyone who promises any investment advice (other than "here's one way to calculate possibilities, review the math for yourself") is doing something suspicious. Buying anything and hoping the value goes up and you can sell it later for a profit always involves risk.

---

[4]Please follow your local regulations about reporting income, paying taxes, and accounting properly. Do good not only for yourself but for your neighbors too.

Instead, focus on value. What can *you* do with the knowledge you have and can gain? What special things can you provide the world? Maybe it's artisanal honey, career counseling, dog-walking, a pop-up makeover stand in a subway, busking with a hammered dulcimer and harmonica, or something else.

Think of it this way–if you had a million Dogecoin and no one wanted to do anything with Dogecoin like pay their rent, buy a delicious pastry, pre-order a book, help a child have heart surgery, or *do* anything by *spending* Dogecoin, what's the value of your holding?

A network of people, computers, calculations, or anything is only as good as the uses it enables. Everyone who uses Dogecoin to accomplish something useful, helpful, entertaining, or just plain *good* makes the value of the network go up. Maybe your favorite pizza bakery doesn't take cryptocurrency now, but maybe they'll see the value in a couple of years as people reading these words keep building!

The more people you encourage to *make transactions*, the more value moves through the network and the more demand there is for the network's value. The price may go up or down or stay the same, but every new adopter means that there's more that you can do with your million Dogecoin–and there are more ways you can receive Dogecoin. Holding for the sake of holding may help you, but if everyone does that, it helps no one. Help yourself and help other people instead. More people spending means more people running nodes, mining blocks, and buying and selling. That makes a stronger network and a stronger Dogecoin.

In the end, worry less about what's going to happen in a possible (but mathematically implausible) universe. Spend your energy to build good things, support other people building good things, and enjoy the ride.

In other words, keep doing only good everyday (Pledge to Do Only Good, pp. 230).

Tip #81   Calculate Block Fill

Dogecoin groups valid transactions into blocks, which miners mine and store to the blockchain in perpetuity. By default, each block has a maximum size of 750 kilobytes (768,000 bytes). The smallest possible transaction requires 240 bytes, so there's a maximum size of about 3200 transactions per block. At one block per minute, on average, the network can currently process about 3200 transactions per minute–about 53 transactions per second.

Those are averages depending on multiple assumptions. Larger transactions (more inputs and/or outputs) need more space. Transactions aren't evenly distributed throughout the day. Some times are busier than others. Even a small change in the average can affect the throughput; an average transaction size of 300 bytes reduces the average transaction count per second to 43, for example.

With additional Dogecoin popularity, the number of transactions per second will grow, so it's useful to know current and historical volume to predict and understand any potential scaling limits. Developers can address these limits before they hold back the network. Individual users can predict the best times to make transactions (when volumes are historically low).

How Big is a Block?

A mined block may be full or empty or in between. If there aren't enough trans-actions to fill a block, the block gets mined anyhow. Similarly, if there are too many transactions to fill a block, remaining unmined transactions stick around until they get mined into a block. If there are consistently too many transactions to fill a block–more than 40 or 50 per second, by the earlier math–transactions will back up and you'll have to wait longer to see yours mined and confirmed.

That's not something to worry about until it happens, but it's useful to be able to calculate this to gauge the popularity of the network and to prepare for this potential popularity. Fortunately, this data is available for every block.

Use the `getblock` RPC command to retrieve block statistics. In this case, look at the `size` field:

```
{.
  "hash": "7cc9d1...",
  "confirmations": 10204,
  "strippedsize": 15865,
  "size": 15865,
  "weight": 63460,
  "height": 2906545,
  "version": 6422787,
  ...
}
```

This block at height 2,906,545 (from September 23, 2019) had a size of 15,865 bytes. Its 47 transactions averaged about 338 bytes per transaction. Given a max-imum size of 768,000 bytes, this block was a little over 2% full. In the past ten minutes of this writing (early May, 2023), the past ten blocks had a minimum of 82 transactions and a maximum of 787.

How Big are the Last 100 Blocks?

Examining one block to gauge network traffic is too limited; sample more over a longer time period. For example, if you're interested in the past couple of hours, you could look at 100 blocks (or 120, for two full hours, on average, but 100 is a rounder number).

This is easy to calculate with Bash or your preferred programming language. If you know the height of the current, most recent block[5], work backwards. Here's a Bash shell loop which can translate to other languages:

```
for i in $( seq 4713820 -1 4713720); do
  HEIGHT=$( dogecoin-cli getblockhash $i );
  SIZE=$( dogecoin-cli getblock $HEIGHT | jq .size );
  echo "$HEIGHT $SIZE"
done
```

Again, jq extracts one field from a JSON output. This code produces a list of two columns of numbers, one the block heights in descending order and the other the size in bytes of each block.

What Can You Do With This?

If you're processing this data in a language more flexible than shell, use the nextblockhash and previousblockhash fields to skip the getblockhash commands in the earlier example.

Calculating the fill percentage of each block is simple math. Calculating the fill percentage over time is also simple math. Collecting this data over a series of time will take more effort, but produces more insight.

You can add more calculations to the block-by-block comparison. While the output of getblock doesn't give you the number of transactions directly, you can calculate that by looking at the transaction data within the block. Similarly, you can examine the time delta between two adjacent blocks and calculate the average transaction processing time. This isn't a *real* number, because the block difficulty is a larger component of mining time, but it's *a* number that might be interesting to plot over time.

Remember, network doesn't want to *fill* each block. It prefers to balance verifying and confirming valid transactions at a reasonable speed, without making anyone wait too long for confirmation or pay too much for processing. Right now there's

---

[5]Use getblockcount.

plenty of available room for additional usage, but increased popularity means developers and miners and node runners will have to plan for new code, consensus, and techniques. Measuring popularity now helps us all to plan the whats and whens we will need.

  # Tip #82   Calculate Block Statistics

The previous tip (Calculate Block Fill, pp. 255) showed how to fetch block data to calculate important details the network itself can't tell you. There's plenty of data you can mine for information, if you have access to that data.

Running a few RPC commands here and there from the console (Command the Core, pp. 42) or `dogecoin-cli` or another RPC client works for small questions. It gets clunky for big questions, such as "what's the average transaction size in the past thousand blocks" or "what's the largest number of transactions in the past two hundred blocks" or even "how full were blocks on average on January 3, 2021". For that, you need a computer[6].

If you're an experienced programmer, you probably already have an idea how to automate this work. If not, have no fear: while you can solve this in multiple ways, this tip shows one good and flexible approach.

Designing Block Statistics Output

What's the essential behavior here? Given a starting point and a total number of blocks to examine, aggregate specific details about those blocks. Some of those details are sums (total number of blocks, total number of transactions, total size). Some are minimums and maximums (highest and lowest sizes, highest and lowest numbers of transactions per block). Others are averages (highest and lowest block fill rates, total average block fill rate). This covers most data types.

What does the program need from the user? Mostly optional data: the height of a block from which to start (if not the current block), the number of blocks to examine (100 is a good default), any RPC credentials (see Authenticate RPC Securely, pp. 73), and perhaps even the maximum block size (if you want to calculate something other than the default of 750 kb).

What should the program produce as output? This could be anything, depending on what you want to do with the data, but structured data is both readable by humans and consumable by other programs, so JSON is a reasonable default.

---

[6]The digital computer kind, not the person who used computing machines back in the 1940s.

Block Statistics Code

Here's some Python you can run, if you have the right libraries installed and have access to a Core node with authentication configured (see Enhance RPC Calls, pp. 80 for the pattern used here in a different language this time):

```python
#!/usr/bin/env python3

from dogecoinrpc.connection import DogecoinConnection
from pathlib import Path
from sys import maxsize
from xdg_base_dirs import \
    xdg_config_home, xdg_config_dirs, xdg_data_home

import simplejson
import click
```

This first block of code loads several Python libraries:

- `python-dogecoin`, which connects to a Core node to perform RPC commands (here referred to by `dogecoinrpc.connection`)

- `pathlib`, which allows manipulation of file and directory paths

- `xdg_base_dirs`, which helps find platform-specific directory locations, used to find the RPC authorization configuration file here

- `simplejson`, which offers more flexible JSON manipulation than than the builtin `json` library

- `click`, which processes command-line arguments

The use of `python-dogecoin` in particular is awkward here; its authorization scheme is different from the scheme expressed in other tips, so your author worked around the library's preferred approach. Use whatever mechanism you prefer; it's easy to swap for something else.

```python
@click.command()
@click.option("--startat",
              default=None,
              help="Height of the block to start",
              type=int)
@click.option("--numblocks",
              default=100,
              help="Number of blocks to analyze",
              type=int)
@click.option("--blockmaxbytes",
```

259

```
                default=768000,
                help="Maximum size of a block in bytes",
                type=int)
@click.option("--user",
                default=None,
                help="Name of the RPC user",
                type=str)
def main(startat, numblocks, blockmaxbytes, user) -> None:
```

The click library handles command-line arguments. Decorators on the main()
function describe the name of flags users can provide, default values, types of
those values, and even help text. Remember to add these variables to main(), or
else you'll get confusing error messages.

```
config_file = Path(xdg_data_home()) / "dogeutils" / "auth.json"
with open(config_file, 'r') as f:
    config = simplejson.load(f)

if user is None:
    print("No user provided; exiting")
    exit(1)

dc = DogecoinConnection(user, config[user], 'localhost', 22555)
```

This next section of code opens the *auth.json* file containing RPC credentials and
attempts to connect via RPC with the authorization code. There's minimal error
checking here; make this code more robust by checking that the configuration file
exists and that any RPC command authorizes correctly.

If you don't run your node on localhost, change the code (or enhance the RPC
configuration to include a hostname and port, or add those as click arguments,
or....).

```
if startat is None:
    startat = dc.getblockcount()

# make the percentages obvious
blockmaxbytes /= 100

res = {
    "startingHeight": startat,
    "minTxCount": maxsize,
    "maxTxCount": 0,
    "totalTxCount": 0,
    "minBlockSize": maxsize,
    "maxBlockSize": 0,
    "minBlockFill": 0,
    "minFillPercent": maxsize,
```

```
    "maxFillPercent": 0,
    "totalBlockSize": 0,
}
```

This code sets up the data structure for the program to print on success. Setting default values makes the output more sensible, no matter what happens. For example, setting the minimum values to sys.maxsize means that subsequent code can unilaterally assign minimum values. Similarly, dividing blockmaxbytes by 100 is effectively the same as multiplying anything *divided* by that value by 100. This approach here saves multiple calculations later[7].

```
hash = dc.getblockhash(startat)

for i in range(1, numblocks):
    block = dc.getblock(hash)
    block_size = block["size"]

    ...

    hash = block["previousblockhash"]
```

Next, the code performs a loop. The code shown elides the body of the loop so that the structure of the loop is more obvious.

Python's range() operator counts from 1 to the number of blocks requested on the command line–100 times by default. Initializing hash outside of the loop for the first iteration means that the loop can assume that value is already set. Reassigning to hash at the end of the loop skips an extra RPC command to get the hash of the block at the next height. Fortunately, previousblockhash is present in the block's data.

If you've read this code closely, you've noticed that this code walks backward from the given block to the destination block. That's deliberate, though you could write the code the other direction if you prefer. Be careful that the nextblockhash exists however.

What's in the loop?

```
res["minBlockSize"] = min(block_size, res["minBlockSize"])
res["maxBlockSize"] = max(block_size, res["maxBlockSize"])
res["totalBlockSize"] += block_size
```

---

[7]Maybe this shortcut is too clever for your tastes. That's fine; make the code as explicit as you prefer.

```
block_tx_count = len(block["tx"])

res["minTxCount"] = min(block_tx_count, res["minTxCount"])
res["maxTxCount"] = max(block_tx_count, res["maxTxCount"])
res["totalTxCount"] += block_tx_count

block_fill_percent = block_size / blockmaxbytes

res["minFillPercent"] = \
    min(block_fill_percent, res["minFillPercent"])
res["maxFillPercent"] = \
    max(block_fill_percent, res["maxFillPercent"])
```

This code runs for every block fetched in the loop. It's basic statistical and aggregation math. Look closely at a couple of things.

First, the content of transactions in a block doesn't matter. Python's `len()` operator returns the number of elements in a list. That provides the count of transactions in the block.

Second, the use of `min()` and `max()` in assignments is the short way to write "if the current value is less than the previously minimum value, use the current value as the new minimum value" and so on for the maximum value. That's why `sys.maxsize` earlier was so useful. It's currently safe to assume that the number of transactions in a block will be less than the value of the largest integer Python wants to handle[8].

```
res["averageSize"] = res["totalBlockSize"] / numblocks
res["averageTxCount"] = res["totalTxCount"] / numblocks
res["averageFillPercent"] =
    res["totalBlockSize"] / (numblocks * blockmaxbytes)

print(simplejson.dumps(res, indent=2))
```

This penultimate code takes place *after* the final loop iteration has concluded. With all of the data from every block aggregated together, we know enough to calculate aggregated statistics for everything as a whole. That final calculation is the most interesting, because the denominator has to be the total number of bytes *available* in every block. You could count this in other ways, but this approach is explicit.

The `print()` command uses `simplejson` to transform the data structure `res` into text. `indent=2` formats the JSON in a way that's easy for people to read (not all jumbled together into a single line).

---

[8]Python *can* handle larger numbers, but `maxsize` is big enough that everything encountered here will be lesser.

```
if __name__ == '__main__':
    main()
```

Finally, this program ends defensively such that you could make it into a Python library if you want.

What Can You Do With This?

This code is more verbose than the previous Bash code, but it's more flexible. It's easier to modify to add more calculations, and the output is fuller and more practical. Given access to a speedy Core node, the code runs in linear time with the number of blocks you want to process.

That overhead might be frustrating if you example lots of old blocks or a lot of blocks or the same blocks repeatedly. The algorithmic complexity of this code is that you have to issue at least one RPC command for each block you want to inspect, which means round trips over a networking call.

Could that be improved? Absolutely–if you can fetch statistics for each block once and only once even if you process it more than once. That's another tip!

  Tip #83   Export Block Metadata

Blocks contain a lot of data beyond transactions. Other tips (such as Calculate Block Statistics, pp. 258) show what you can do with block data itself: calculate how full blocks are, aggregate data about blocks, et cetera. Those tips all rely on your ability to connect to a running node and get data from that node.

Sometimes that's feasible. If your needs are small or you're querying a small amount of data, this can even be fast. Other times, you may not have a node available. You may want to analyze historical data. You may want to explore a week, month, or year's worth of blocks. You may even be on a long airplane or train ride and want to do some hacking[9].

Sometimes apparent disadvantages can be solutions in disguise, if you reframe the problem. If the problem is "Dogecoin Core isn't really designed for bulk aggregate analysis of block metadata", one solution is to export the data to make that analysis easier.

You've probably used one of the potential solutions more than once today without realizing it.

---

[9] Your author once started writing payment control software for a pinball arcade on an airplane.

263

Storing Normalized Data

You may have already used a *relational database* to store, retrieve, or manipulate data. In this system, something called a *table* defines a structure for *rows* of data. Perhaps this is a name, a date, and a high score for a Lord of the Rings pinball machine or the Crystal Castles arcade game. Every entry in the table is a row containing those three pieces of data, interpreted as a name, a date, and a numeric score.

You can ask questions such as "Who has the highest score of all time?" or "What are all of Unky c's scores?" or "What are the highest scores per machine per week?" The normalized structure of the data and the mechanics of the storage system make these arbitrary, ad hoc queries possible.

Does that sound useful?

SQLite is Ubiquitous

Several good open source relational databases come to mind: PostgreSQL, MariaDB, MySQL, and SQLite. You may have a favorite. This example uses SQLite, because it's lightweight, requires almost no setup and maintenance, and runs almost anywhere. You're probably already using it on your phone, tablet, laptop, and other devices without even knowing it.

A Crypto Connection

Besides that, Bitcoin appears to be ready to adopt SQLite as the storage format for wallet data, which means that Dogecoin will also adopt this feature. Watch for updates!

If you don't already have a program called `sqlite3` installed on your computer, visit the SQLite homepage[10] to download and install your own copy. This program is available using operating system-specific package managers such as Homebrew, Chocolatey, apt, and yum.

After installing SQLite, it's time to answer two questions. What's the shape of data (the table schema definition, in relational database terms)? How do you get data from a Dogecoin Core node into SQLite? First things first.

---

[10]As of this writing, https://sqlite.org/–but always verify!

Defining a Block Metadata Schema

What do you want to know about blocks? Height and hash, for sure. With that information, you can identify a block uniquely on the Dogecoin blockchain. As well, keeping the hash and not just the height helps you verify that the information you have for a block matches other information you can find online. Hold this thought.

Other tips discuss block difficulty, number of transactions, and alluded to the time of a block's mining. That's several fields, so it's a good place to start. If you want other fields available from blocks, you can always modify the code presented here.

To define a table in SQLite, give it a name, name each column of data (a row is a series of columns), and provide detail about the type of data in each column. Although SQLite mostly doesn't care about data types, it's still a good habit to think about what's what so you can avoid surprises later.

Create a file named *schema.sql*, containing:

```
CREATE TABLE blockstats (
    height     INTEGER NOT NULL,
    epochtime  INTEGER NOT NULL,
    hash       CHARACTER(64) NOT NULL,
    size       INTEGER NOT NULL,
    txcount    INTEGER NOT NULL,
    difficulty FLOAT NOT NULL
);

CREATE INDEX blockstats_height_idx    ON blockstats (height);
CREATE INDEX blockstats_epochtime_idx ON blockstats (epochtime);
```

These six columns correspond to the six metadata fields identified earlier. Height, block time (called "epoch" time, because Dogecoin stores them as seconds since the Unix epoch of January 1, 1970), size, and count of transactions in the block are all integers. The hash of the block is a text field exactly 64 characters long. The difficulty is a floating point value: it has a decimal point and the precision of that number is important. All of these values must be present in every row; none of them can have a null ("not present") value.

This CREATE TABLE statement tells SQLite to create a table named blockstats and prepares that table to hold zero or more rows of data with information that matches the types of data expected.

The two CREATE INDEX statements are optional, but you'll find them useful. They tell SQLite to track additional information about the height and epochtime columns so that any queries written to explore data in this table can use either field to look up relevant data faster. For example, if you wanted to find the block at height 2,345,679, SQLite can look in the `blockstats_height_idx` index for the location of the relevant row. Without that index, SQLite would have to search the table to find that block–on average, it would have to look through half the rows in the table to find it.

Indexes are Like Spice

No, indexes don't make Guild navigators effective, but they also aren't useful on *every* column. Every index takes up disk space and makes it slower to insert data into a table. That may not matter for this tip.

Also remember that a database can use, in general, only one index per query per table. There's little value in adding extra indexes, if every query of this table includes either the block's height or the time of its mining. These two indexes will cover (pardon the pun) everything relevant.

From the command line, initialize a new database and create that table:

```
$ sqlite3 blockstats.sqlite < schema.sql
```

If all goes well, you now have a single file called *blockstats.sqlite* ready to store new data!

With an empty database, you need data. You can populate this table in several ways, including:

- connect to SQLite from your favorite programming language

- write SQL INSERT statements from your favorite programming language

- export a non-SQL data file from your favorite programming language

All of these approaches work. The latter might be easiest to explain, and it shows off the intermediate data. In short, loop through all of the blocks a Core node has, grab each block's relevant data, and write that data, one line at a time, to a CSV file (seen in Find All Received Addresses, pp. 142). Open this file in a spreadsheet or skim it with a text editor; SQLite can also import it natively.

The Python code to export data from multiple blocks looks a lot like the code used to aggregate data from multiple blocks.

266

```
#!/usr/bin/env python3

from dogecoinrpc.connection import DogecoinConnection
from pathlib import Path
from xdg_base_dirs \
    import xdg_config_home, xdg_config_dirs, xdg_data_home

import csv
import click

@click.command()
@click.option("--startat",
                default=1,
                help="Height of the block to start",
                type=int)
@click.option("--numblocks",
                default=100,
                help="Number of blocks to export",
                type=int)
@click.option("--user",
                default=None,
                help="Name of the RPC user",
                type=str)
@click.option("--outfile",
                default="blockexport.csv",
                help="Name of the output file to write",
                type=str)
def main(startat, numblocks, user, outfile) -> None:
```

The `click` command-line arguments here > are slightly different. With no default values provided (except a username), this code exports data for the first hundred blocks. You can start at any arbitrary height and process any arbitrary number of blocks. Be sure to add error checking to see if you've exceeded the bounds of what your node can process!.

This code writes its output to a file called *blockexport.csv* by default[11].

```
config_file = Path(xdg_data_home()) / "dogeutils" / "auth.json"
with open(config_file, 'r') as f:
    config = simplejson.load(f)

if user is None:
    print("No user provided; exiting")
    exit(1)

dc = DogecoinConnection(user, config[user], 'localhost', 22555)
```

---

[11]To make this program more robust, check that the requested destination file exists before attempting to write to it.

This code is the same as in the previous tip.

```
lastblock = startat + numblocks - 1

hash = dc.getblockhash(startat)
```

Unlike the previous code, this program works from the earliest block to the latest requested block. By calculating `lastblock`–the height of the final block to process–the loop in the next code snippet is easy to reason about.

```
with open(outfile, "w") as f:
    writer = csv.writer(f)
    want_values = \
        ["hash", "height", "time", "size", "tx", "difficulty"]
```

This section of code uses Python's `csv` library to wrap a file opened for writing (`"w"`). You'll see this used shortly to write every line of data. This wrapper formats every line appropriately for the CSV file. It doesn't matter with the data exported here, but if there were any special characters, this library would handle them without making a confusing or broken CSV file.

```
for i in range(startat, lastblock):
    block = dc.getblock(hash)
    block_values = [
        block["height"],
        block["time"],
        block["hash"],
        block["size"],
        len(block["tx"]),
        block["difficulty"]
    ]
    writer.writerow(block_values)
    hash = block.get("nextblockhash", None)

    if hash is None:
        break
```

Finally, for every block height in the range from the starting block to the final block desired, this code fetches the block at that hash (`hash` initialized before the loop, as before), extracts several fields from the `block`, stores them in a Python list, then writes that list to the CSV file through the CSV `writer`. Then the code resets `hash` with the value of the next block.

If there's no next block, the loop ends. The loop also ends once it reaches the height of the final block.

```
if __name__ == '__main__':
    main()
```

As always, this idiomatic Python code makes it easier to turn the entire program into a library for use elsewhere. Run the program with:

```
$ python3 exportblocks.py --user lisa --numblocks 4800000
```

As of May 2023, there were 4.7 million blocks on the chain, so this code will run for a couple of hours and export all of those blocks. Change the number as suits your needs.

Import Exports

With the SQLite database created, a schema initialized, and a bunch of blocks analyzed and stored in a CSV file, it's time to put everything together.

```
$ sqlite3 blockstats.sqlite
sqlite> .mode csv
sqlite> .import blockexport.csv blockstats
```

Your computer will process everything, and you'll end up with a `blockstats` table full of block statistics data!

What Can You Do With This?

Any analysis you can perform with SQL is available on this data now. For example, you can:

- Count all of the blocks mined on any specific day

- Get the average number of transactions within a range of heights

- Show the number of seconds between any two blocks

- ... and more

You also now have a source of data–disconnected from your node–to use to verify anyone else's claims. Suppose you don't trust this author's repeated refrain that there are about 60 seconds between blocks on the Dogecoin network[12] You can calculate this yourself!

---

[12] You're in good company; core developer Patrick Lodder also disagrees.

Suppose you disagree that the most congested blocks happen at 3:00 UTC. You can discover this yourself!

Suppose someone claims that the hash of block 123,457 starts with the improbable prefix CAFED00D. You can check that yourself!

Exporting and importing this data takes some work and storage space–not to mention any ongoing efforts to keep it up to date–but it provides you an invaluable source of information and a source of truth you can maintain yourself.

  Tip #84   Analyze Block Metadata

With block metadata in a database (see Export Block Metadata, pp. 263), you can ask arbitrary questions about block data. The database provides *structure* for the data to allow you to analyze the blockchain however you can imagine.

With SQLite available, open a command-line window (a terminal app on Mac OS, something like PowerShell on Windows, and a terminal app on Linux or another Unix-like system).

With your *blockstats.sqlite* database, type `sqlite3 blockstats.sqlite`. At the new prompt, you can write SQL[13] A SQL query is a request to the database to read, write, update, or delete data.

SQL and relational databases build on an entire field of knowledge with a lot of theory and practice behind it, but with some basic understanding, you can answer several questions with the power of a database.

Count Blocks by Day

How many blocks get mined in a day? You could take the author's word that a block gets mined, on average, every 60 seconds. That suggests about 1440 (60 * 24) blocks per day. Is any day actually average?

What does it mean to count blocks by day? Based on the structure of this data, each block has its mining time as measured in epoch seconds. That means a query of the `blockstats` table to count all blocks per day must group blocks into days somehow.

The SQL query for this reads *exactly* like that, or at least exactly-ish enough. Type this query:

---

[13]Structured Query Language, pronounced sequel or ess-cue-ell or squirrel or whatever you prefer, as Donald Chamberlin once told the author.

```
sqlite> SELECT COUNT(*), DATE(DATETIME(epochtime, 'unixepoch'))
FROM blockstats
WHERE DATE(DATETIME(epochtime, 'unixepoch')) >= '2023-04-01'
GROUP BY 2
ORDER BY 2;

1353|2023-04-01
1328|2023-04-02
1347|2023-04-03
1335|2023-04-04
1346|2023-04-05
1354|2023-04-06
1349|2023-04-07
1354|2023-04-08
...
```

The indentation doesn't matter for the database, but it makes reading the query easier. Look at the first word of each line.

SELECT tells the database that this is a read operation. It won't modify or delete data. It reads some data based on certain conditions. The rest of the line describes two columns of data to see in the output (you can see them in the output). Hold that thought.

FROM tells the database the source of the data. There's only one table in this database so far, but it's good to be specific.

WHERE is optional. Here it filters a subset of the data.

GROUP BY tells the database to collect a bunch of data into buckets that make sense based on other conditions. Here 2 refers to the second column (columns follow the SELECT keyword).

ORDER BY is optional. It tells the database to sort the output by one specific column, again the second column.

What are these columns?

COUNT(*) is an intrinsic operator that tells the database to count things. In technical terms, it's an aggregate operator, which means that, on its own, no matter how much data is in the table, you'll get one result. If you wrote SELECT COUNT(*) FROM blockstats;, you'd get a count of the total number of rows in the table.

DATE(DATETIME(epochtime, 'unixepoch')) is kind of a mess[14] You may recognize the epochtime column from when you inserted data into the database. This expression has two conversions. DATETIME(epochtime, 'unixepoch')

---

[14]The kind that makes your author wonder if he made your life more difficult by not converting the exported data.

asks SQLite to take the value of `epochtime` and convert it from Unix epoch seconds into a date and time value, like "high noon on April 22", but more mathy. That's a lot more helpful for visual display, because 2023-04-08 makes sense to humans.

The outer function, `DATE(...)` truncates a date and time value into only a date, which is useful because the point of this query is to count how many blocks get mined per *day*, not per minute or second.

You see the calculation for that second column repeated after the `WHERE` keyword; converting the `epochtime` column to a date again makes the comparison against April 1, 2023 work correctly. With that filter in place, SQLite will examine only those rows representing blocks mined at or after 2023-04-01.

That gives you enough information to understand `GROUP BY` now. Remember that `COUNT(*)` produces a single value. That means the first of the two selected columns is an aggregate. The other column isn't; if you wrote a query `SELECT DATE(...) FROM blockstats`, you'd get one result for every row in the table. SQLite can't reconcile the different cardinality of these two columns on its own, so it's up to you to disambiguate. You need *more* count rows and/or *fewer* date rows.

`GROUP BY 2` means "put the contents of each row into a bucket based on its date". The output is equivalent to writing a bunch of `SELECT COUNT(*) FROM blockstats WHERE DATE(...) = '2023-04-01'` queries and pasting the output together yourself (but make the computer do the work).

Count Transactions by Day

Now you may be thinking "If the author's wrong about 1440 blocks per day, maybe he's right about the number of transactions per day". Fortunately for your author's ego, at the time of this writing the number of transactions per day spiked.

Try this query.

```
sqlite> SELECT SUM(txcount),
        DATE(DATETIME(epochtime, 'unixepoch'))
FROM blockstats
WHERE date(datetime(epochtime, 'unixepoch')) >= '2023-05-07'
GROUP BY 2
ORDER BY 2;
18681|2023-05-07
20444|2023-05-08
41033|2023-05-09
49918|2023-05-10
457854|2023-05-11
398786|2023-05-12
629570|2023-05-13
```

Although you've seen similar queries, SUM(txcount) is different. The txcount column contains the number of transactions in each block. Like epochtime, it's a scalar (non-aggregate) column. For GROUP BY to do the right thing here, the number of transactions column has to become an aggregate column. Fortunately, SUM(...) aggregates all of the rows bucketed by a single day and adds their values together.

What Can You Do With This?

Now you can modify or adapt these queries to count or sum any data you've collected over any timeframe. For example, to see the number of transactions by hour, change the selected column from DATE to another SQLite expression that lets you group by day and hour (or only hour, if you're looking at a single day).

If you add other data to this database when you export it from your node, you can count and group that data as well.

What if you want to calculate things like averages? Read on!

  Tip #85   Average Block Metadata

With block metadata in a database (see Export Block Metadata, pp. 263), you can query and analyze the data. You've seen some of this in Analyze Block Metadata, pp. 270. You can ask even more complex questions and perform more powerful manipulations on the data. The more SQL you understand, the more options you have.

Alias Common Columns

The previous tip reused the awkward manipulation of turning the epochtime column into a date column. Why not let SQLite do that for you? While you *could* reimport all of the data or create a new table or add a column to calculate the date, you can easily change the way you *view* the data.

A view in SQL, at least as practiced by SQLite, resembles a query you can reuse in other queries. Alternately you might think of it as a way to make something that looks like a table but which contains aliases for other data. Wouldn't it be nice to write SELECT date, txcount... instead of all that other complexity?

Save this file as *create_dates_view.sql*:

```
CREATE VIEW IF NOT EXISTS blockstats_dates
AS
SELECT
```

```
      height,
      date(datetime(epochtime, 'unixepoch')) AS date,
      epochtime,
      hash,
      size,
      txcount,
      difficulty
   FROM blockstats;
```

Run it with `sqlite3 < create_dates_view.sql blockstats.sqlite`, or type that code directly into a SQLite prompt.

That code does exactly what it implies. If there's no existing view in the database with the name `blockstats_dates`, it creates that view. That view has columns from the `blockstats` table as well as a new column, `date`, which represents the block's date.

To list all of the heights and hashes from blocks mined on your goddaughter's birthday, write:

```
SELECT date, height, hash
FROM blockstats_date
WHERE date = '2023-04-22';
```

Views like this are generally cheap and, if you keep them to SQL statements like this, reasonably easy to manage, so feel free to manipulate the data to make subsequent queries easier to write without having to modify existing data.

Average Difficulty by Day

With that scaffolding in place, it's easier to answer questions like "What's the average mining difficulty every day in April 2023?" If you've studied the previous tip, you probably only need the hint that SQLite has a function called `AVG` to calculate the arithmetic mean of a data set.

```
sqlite> SELECT DATE, AVG(difficulty)
FROM blockstats_dates
WHERE date >= '2023-04-01'
AND    date <  '2023-05-01'
GROUP BY 1
ORDER BY 1;
2023-04-01|10246432.0898101
2023-04-02|10448319.7542918
2023-04-03|9996381.92059544
2023-04-04|10325399.6426264
2023-04-05|10596347.1752015
2023-04-06|10424084.9092164
```

```
2023-04-07|10825771.5585102
2023-04-08|10778853.3159625
2023-04-09|11334822.9295326
2023-04-10|10716047.2139055
2023-04-11|10619734.8399633
2023-04-12|10064203.9079213
2023-04-13|10254175.2447795
2023-04-14|10262071.1208702
2023-04-15|11064251.7274149
2023-04-16|10968131.3985921
2023-04-17|10617461.8932077
2023-04-18|10170954.1527323
2023-04-19|10205503.8287437
2023-04-20|10554817.552975
2023-04-21|11036352.148357
2023-04-22|11018358.3450682
2023-04-23|10658572.8425474
2023-04-24|10478176.3345837
2023-04-25|10347352.0460158
2023-04-26|10333477.0577914
2023-04-27|10159316.8870308
2023-04-28|9977940.50760047
2023-04-29|10400965.385681
2023-04-30|10524277.7201581
```

This query is easier to read because of the `blockstats_dates` view, especially because the `WHERE` clause refers to the `date` column twice. It's totally okay to refer to a column in the `WHERE` clause multiple times like this[15].

Average Block Fill Rate

The query to see the average block fill rate over time is also straightforward:

```
sqlite> SELECT date, AVG(size/7680.00) AS fill_rate
FROM blockstats_dates
WHERE date BETWEEN '2023-05-01' AND '2023-05-14'
GROUP BY 1
ORDER BY 1;
2023-05-01|1.08704812885802
2023-05-02|1.23228844227968
2023-05-03|1.12467042223046
2023-05-04|1.07806550082781
2023-05-05|1.16125461115143
2023-05-06|1.09360965909091
2023-05-07|1.02771558204468
```

---

[15]You can also write `WHERE date BETWEEN '2023-04-01' AND '2023-04-30'`, but notice that the range of dates searched includes both the low and high values.

```
2023-05-08|1.1424784402819
2023-05-09|1.83941647376543
2023-05-10|2.04727018371627
2023-05-11|16.0814344697665
2023-05-12|14.0762417192917
2023-05-13|21.7191201513899
2023-05-14|18.8563407512626
```

This query has two interesting features. First, it divides the `size` column by 7680.00. Why the extra zeroes? Mathematically that makes no difference, *except* that including them tells SQLite to keep any fractional component of the remainder for display. Without that precision, SQLite might throw away information[16]. Second, there are 768,000 bytes in 750kb, but we want to display the fill rate as a percentage, so we'd have to multiply the results by 100.00 to get percentages. Because your author is very cautious about minimizing floating-point math where possible, he's consolidated the calculations into a single value, so that SQLite will reduce compounding errors in mathematically-equivalent additional calculations.

Find the Maximum and Minimum Fill Rate

Average fill rates are interesting, but extraordinary blocks are most interesting. If the average block is 5% full but you really want your transaction to confirm quickly, and if the last 10 blocks have been 99% full, the average is less interesting.

Here's how to calculate the maximum and minimum fill rate by day. First, why not make another view (or add to the previous view) to make a `fill_rate` column?

```
CREATE VIEW IF NOT EXISTS blockstats_fill_dates
AS
SELECT
    height,
    date(datetime(epochtime, 'unixepoch')) AS date,
    (size / 7680.00) AS fill_rate,
    epochtime,
    hash,
    size,
    txcount,
    difficulty
FROM blockstats;
```

Use the SQL `MAX()` and `MIN()` functions to find the highest and lowest values for a column, respectively. These are aggregate functions, so to group them by day, `GROUP BY` the date column:

---

[16]Try it without the extra digits and see what it does for you!

```
sqlite> SELECT date, MAX(fill_rate), MIN(fill_rate)
FROM blockstats_fill_dates
WHERE date BETWEEN '2023-05-01' AND '2023-05-14'
GROUP BY 1
ORDER BY 1;
2023-05-01|57.7990885416667|0.0705729166666667
2023-05-02|29.693359375|0.0798177083333333
2023-05-03|72.4002604166667|0.068359375
2023-05-04|28.799609375|0.0682291666666667
2023-05-05|23.9272135416667|0.0846354166666667
2023-05-06|27.1270833333333|0.068359375
2023-05-07|19.9759114583333|0.0798177083333333
2023-05-08|56.0169270833333|0.083203125
2023-05-09|62.9592447916667|0.0923177083333333
2023-05-10|46.2252604166667|0.0765625
2023-05-11|129.995052083333|0.09296875
2023-05-12|97.655078125|0.0725260416666667
2023-05-13|97.66328125|0.068359375
2023-05-14|97.6610677083333|0.09296875
```

A block was almost 130% full on May 11! This may indicate a failure in data extraction, a failure in the calculation, or a failure in understanding block size limits. Whatever the cause, it's a good reminder to verify things for yourself; always be willing to challenge someone else's assertions and check your sources.

What Can You Do With This?

You can take this in several directions: start making graphs about the health of the network over time, plan your own transactions to be gentle on the network (or maximize your throughput), perform advocacy for people running nodes or using Dogecoin as a currency, and more.

Data analysis skills are useful in many lines of work and areas of research. If you already have those skills, apply them here to help developers and node runners decide where to focus further attention. If you're looking to develop those skills, this is real data available to anyone where you can ask and answer all sorts of interesting questions.

   Tip #86   Speed Up Analysis

Depending on the speed and workload of your computer, all of this data analysis (see Analyze Block Metadata, pp. 270 and Average Block Metadata, pp. 273) can be somewhere between "pretty fast" and "slow". Although computers are on average much faster than people, they can be slow because they aren't smart.

If you know there are currently 4.7 million blocks and you're looking for what happened in block 4.69 million, and if you know you've exported blocks in ascending height order, you don't have to read through the heights of 4.69 million blocks to find the one you're looking for. You'll start somewhere near the end and work your way backwards to narrow down your search.

A computer isn't that smart. Run a query to look up a specific block's date based on its height and SQLite will think for too long. Try it and see:

```
SELECT height, date FROM blockstats_dates WHERE height = 4,698,765;
```

Index Your Data

What's the smart thing for a database to do? That depends on what it knows. If it knows nothing about the data, then it must check every row until it finds the desired `height`.

If you've exported 100 block heights (1 to 100, for this example) and you search for each block height, looking for the block at height 1 means one lookup. Looking for the block at height 100 means 100 lookups. Add up all of those lookups and divide by 100 to get the average number of lookups: 100 divided by 2, or 50. With a thousand blocks, the average number of lookups is 500–or 1000 divided by 2.

You can see where this is going.

To make your database queries faster, add an *index* to a column or columns. Essentially this is a way of saying "I'm going to ask you for information based on this data, so track that data in a way that may let you do smart things."

Index Your Height

For example, to look up block information based on their heights faster, type this into your SQLite prompt:

```
CREATE INDEX blockstats_height_idx ON blockstats(height);
```

This tells SQLite to attach an index to the table `blockstats` on the `height` column, so that any query you write that uses `height` in the `WHERE` clause has a chance of using the index rather than looking at, on average, half of the rows in the table[17].

---

[17]How did your author know how to do this? Years of experience. Get a head start by reading about query plan explanations at https://www.sqlite.org/eqp.html.

Index Your Date

What if you're not querying the `blockstats` table but instead one of the views from a previous tip? What if you're not looking for height but by date?

Good news:

```
CREATE INDEX blockstats_date_day_idx
ON blockstats (date(datetime(epochtime, 'unixepoch')));
```

Instead of indexing a column by itself (`epochtime` here), this adds an index on *functions* performed on the column. This speeds up any query that uses that specific application of functions on `epochtime` in a WHERE clause *as well as* any views defined that turn that function into something that looks like a column.

In other words, if you suffered through slow queries reading the previous tips, your author apologizes a little bit but deliberately chose these examples to make a deeper point[18].

What Can You Do With This?

"Look, Unky c, I wanted to read a book about dog-themed money, and you made me install a relational database you said I already had installed on my system, and now you're teaching me about computer science. I care about dogs and money and we're well beyond that!"

Okay, that's a fair reaction.

Did you ever wonder why you have a wallet? With watch-only addresses (see Watch Wallet Addresses, pp. 157), you can keep track of transactions even if they don't belong to you—but why do you have to go through extra work just to register them with a Dogecoin Core? What's the use of `txindex` (Index All Transactions, pp. 192) anyway?

With the full blockchain approaching 100 GB in size when stored on disk, it'd be incredibly inefficient to look up anything without an index. As written before, the best case scenario is something like scanning 50 GB of data on average for every question you have about the blockchain. With every transaction, things get slower.

There are only three solutions, and one is no good right now:

- wait for computers to get faster

- store less data

---

[18] If you've skipped around through the book, good for you!

- store data structured for quicker lookups

Wallets hit point two. Indexes hit points two and three. When you add an address to your wallet and rescan blocks, you're taking wallet-specific information out of the entire blockchain and storing it elsewhere for faster wallet-specific operations. When you index all transactions, you're taking transaction-specific information out of the entire blockchain and storing it in a way so that the Core can look things up faster.

Everything from blocks to transactions to statistics about blocks is just a big batch of data stored on a computer somehow. It's up to us humans to attach meaning to it. Sometimes that meaning is us sending it around to other people. Other times, it's us telling computers how to structure the data for improved meaning.

Now you know some of the principles behind other operations done on your behalf or done at your request in Dogecoin Core and other software *and* you have tools to make your own analyses better/faster/smarter.

Tip #87　Estimate Dogecoin's Current Conversion Value

If Dogecoin is money, you should be able to exchange it for other valuable things, such as goods, services, and other forms of currency. Your consulting business where you speed up SQL queries (see Speed Up Analysis, pp. 277) for 1000 Dogecoin an hour is pretty cool, but when it comes time to visit the local farmer's market and buy a flat of Hood Strawberries during the three weeks in June when they're available, you might want USD or at least CAD on hand.

Or you might have in your mind a conversion where one quarter equals one song played in your Dogecoin jukebox (see Control Your Jukebox, pp. 221), but what exactly does "one quarter" mean on the blockchain?

You can find out and then keep finding out.

Conversion Rate APIs

Several exchanges and crypto-related services provide APIs to report the current exchange rate of any coin in one or more fiat currencies. For example, you could ask "What would I pay for one Bitcoin in UK Pounds" or "How much Litecoin can I get for 1 Rupee".

The intent behind many of these APIs is to charge developers and businesses building crypto-related products and services for access, so many of these systems want you to sign up and subscribe to paid services. This can be a good thing in some ways, but for example purposes, this tip uses `coinranking.com` because it has a free API with generous limits for testing.

> **Time Marches On**
>
> Depending on how much time has elapsed between the writing of this tip (August 2024) and your reading of this tip (insert local time and date here), this example may no longer work or the URL may no longer be useful, available, or accurate. That's okay; ask around, search for alternatives, and be prepared to switch your code. The concepts still apply even if the details may change.

As of this writing, Coinranking has two API endpoints you need to use and another you might want to use: get list of coins[19], get coin price[20] and get list of reference currencies list[21].

The most important call is get coin price. You need a coin's UUID, which you can retrieve from "get list of coins". Use the tools `curl` and `jq` from the command line:

```
$ curl https://api.coinranking.com/v2/coins > coins
$ jq '.data.coins[] | select(.name == "Dogecoin").uuid' < coins
"a91GCGd_u96cF"
```

Note that you could swap `.uuid` for `.coins` and get the price in an all-in-one query, but this two-step process is kinder to the server for anonymous use.

Given the UUID for Dogecoin (skim the *coins* file to convince yourself this is correct), the "get coin price" query is:

```
$ curl -s \
  https://api.coinranking.com/v2/coin/a91GCGd_u96cF/price | \
  jq .data.price
"0.06701734282680077"
```

---

[19] See https://developers.coinranking.com/api/documentation/coins#get-list-of-coins.

[20] See https://developers.coinranking.com/api/documentation/coins#get-coin-price.

[21] See https://developers.coinranking.com/api/documentation/reference_currencies.

The `-s` switch to `curl` silences extraneous output. Be careful that this pipeline of commands doesn't handle errors, such as "your network is not connected", "you need an API key", "the server rate-limited your requests", or "this API has gone away". Add whatever error-checking you think you might need.

The default currency value is US Dollars. If you need something else, look for the currency UUID from the reference currencies list:

```
$ curl https://api.coinranking.com/v2/reference-currencies |
  jq '.data.currencies[] | select(.name == "Turkish Lira").uuid'
"Dk5T6J0UtCUA"
```

...then provide the currency UUID to the price call[22]:

```
$ export API=https://api.coinranking.com/v2/coin
$ export CURCODE="referenceCurrencyUuid=Dk5T6J0UtCUA"
$ curl -s "$API/a91GCGd_u96cF/price?$CURCODE" | jq .data.price
"1.70040047930945557576"
```

That's 1.7 lira (Turkish or Cyprian), give or take, per Dogecoin.

With this query example in place, you can automate this system to:

- Run this job hourly, or on whatever schedule works for you

- Update a value in a database with the current price

- Send you a notification when prices rise or fall outside of specific thresholds

Again, use this only if the *current* conversion rate between your local currency and Dogecoin is meaningful for specific transaction purposes such as pricing inputs and outputs appropriately. Otherwise you run the risk of worrying too much about something that's supposed to be fun.

Understand the Risks

Whenever you rely on external data you don't control yourself, you face at least two risks. First, the data could be inaccurate for any reason: it's been hijacked, it's under the control of malicious actors, it's out of date, it's calculated incorrectly, et cetera. Depending on your needs, you might want to verify these values against multiple independent sources. In a case like this, the numbers probably

---

[22]The linebreaking in this example is for display purposes; you can run this command as a single line.

won't match exactly, but you might expect them to converge over time on median values no more than one sigma apart.

If that doesn't happen, ring the alarm bell!

The second risk of relying on an external data source is that data source's longevity. What happens if the API goes away? If your work depends on fresh and accurate data, then establishing a business relationship with an entity with a contract and obligations can reduce your risk.

Another, more oblique risk is your personal appetite to watch the value of your holdings fluctuate on a regular basis. If you've invested hours and dollars into building up your system and a one-penny move downward is depressing and disheartening and a one-penny move upward is euphoric, think about what you want to accomplish in the long term and try to find a way to measure your success over time.

If you need to cash out every day to mitigate your risk, that's fine. If you're holding coins because you don't need them to pay your bills right now, spend your time making sure your other finances support your choices.

In the end, we're all just swapping friendly dog-themed ideas based around large numbers a bunch of computers are swapping. To paraphrase Charlie Papazian: "Relax. Don't worry. Have a Dogecoin."

Tip #88   Set Your Own Price

You're selling a good or a service, maybe blueberry lemonade or rocky road ice cream. Perhaps it's a book or consulting services or credits in your pinball arcade. Unless you will only ever use Dogecoin as your currency throughout your business, you need to meet your costs somehow. That means knowing how much Dogecoin you need to charge. With the current conversion value to your favorite fiat currency (set Estimate Dogecoin's Current Conversion Value, pp. 280), you know enough to set your own price.

Fill Up Your Price Threshold

Suppose you're selling an ebook. Your cost calculation suggests that the ideal price is $15 USD per copy. At that, you've made back your investment and can turn any profits into hardware upgrades to write the next edition of the book. You have a website where you sell this ebook. Of course you take Dogecoin[23], but the value

---

[23] And in this case, Pepecoin, thanks to the advocacy of that network's creator.

of Dogecoin fluctuates from day to day.

If you'd priced the book at Dogecoin's all time high of around $0.70 USD, you'd price the book at 21.5 Dogecoin per download. If you'd priced it at a low of around $0.06, you'd charge 250 Dogecoin per download. That's a big gap! Floating the price you charge based on the current conversion rate can help.

The simple and direct approach is to set your price threshold and divide based on the current conversion rate, but where's the fun in that? Simpler numbers seem kinder and easier to manage. After all, if you take this approach and apply it to playing music on a jukebox or a round of Dig-Dug, do you really want to say "insert 3.5663 Dogecoin to continue"?

It's better to create a formula to produce a nicer, rounder number. Maybe one decimal point is enough. This JavaScript/Node code does that:

```
#!/usr/bin/env node

function calculateDogecoinThreshold(currentRate, desiredPrice) {
    const naiveThreshold = desiredPrice / currentRate;
    return Math.ceil(naiveThreshold * 10) / 10;
}

const [ currentRate, desiredPrice ] = process.argv.slice(2);

const suggestedPrice =
    calculateDogecoinThreshold(currentRate, desiredPrice);

console.log(
    suggestedPrice +
    " Doge at " +
    currentRate +
    " is " +
    suggestedPrice * currentRate +
    " (" + desiredPrice + ")"
);
```

The `calculateDogecoinThreshold` function takes the current conversion rate and your desired price. For example, if Dogecoin is currently worth $0.068 USD and you want to sell each ebook for $15.00 USD, run `node calcPrice.js 0.068 15.00` to get this output:

```
$ node calcPrice.js 0.068 15.00
220.6 Doge at 0.068 is 15.0008 (15.00)
```

How does this work? The calculation is straightforward; divide your target price by the current conversion rate. This will give you a big number. Then multiply the value by 10 (representing a single decimal place), use Node's built-in `Math.ceil` function to strip off the remaining decimal places, and finally divide by 10 again. This will give you a nicer, rounder number.

You can get more creative with this. 220.6 Doge is close enough to 220 that you could rearrange the value down to a whole number. Alternately, if the conversion price were $0.076432, you could round 196.3 Doge up to 200 Doge. Writing that algorithm will take more work, but it's a good exercise if you're working your way through *Structure and Interpretation of Computer Programs*.

Understand the Risks

What can go wrong? Given that this is JavaScript code, you or someone you love might get the temptation to use this as client-side code in a web page. *If* you do that[24], a minimally savvy user can figure out how to set their own price and end up getting liters and liters of cheap lemonade.

If your conversion rate update threshold is too tight, you can end up with people watching the price change frequently as they're trying to buy. In that case, you might prefer a price affinity mechanism tied to the user's session, so that they get a nicer experience[25].

Finally, if you automate all of the mechanisms you use to set prices, you limit your ability to do interesting things such as offering bulk purchase discounts, coupons, or other promotional offers. You may experience these cases when you provide a token system, for example (see Manage Tokens, pp. 311).

  Tip #89   Check Your Parameters

Back in the olden, more innocent times of 2013, there was Bitcoin and then a lot of things happened. A coin named Litecoin began by forking Bitcoin[26]. Then Luckycoin forked Litecoin. Then Dogecoin forked Luckycoin.

All of these coins can trace their lineage back to the original Bitcoin source code.

By now you probably know what makes Dogecoin different: it's a joke that somehow grew out of hand, it has a feature called merged mining that shares hash

---

[24]Don't do that.

[25]Though again, if you store and retrieve the price information to and from the client, the client can change this information.

[26]Someone made a copy of the original source code, made some changes, released the changes, and development continued on the modified version.

power similar coins such as Litecoin, it has a very fast block time at least 10x that of Bitcoin, and it has a culture of fun, generosity, and creativity.

If you're feeling reductive, however, and look at cryptocurrency as a bunch of numbers that computers and nerdy people swap around, then you might ask "What *really* makes any of these cryptocurrencies different?" Good question.

Network Parameters

If you have the hardware for it (see Run a Node, pp. 34), you could run a Bitcoin node, a Litecoin node, and a Dogecoin node all on the same computer. While the messages they send back and forth across the network are all *similar*, each coin has a set of network parameters that prevent them from overlapping.

For example, Bitcoin has a network port of 8333, Litecoin 9333, and Dogecoin 22556. When your node tries to connect to other nodes, it will use the port appropriate for that specific coin. If you try to connect to a Dogecoin node on port 9333, you won't get good results.

> **What's a Network Port?**
>
> Think of a port sort of like a series of mailboxes in an apartment building, where every apartment has a number. If you want to send a letter to someone in apartment 221b, you'd better address it correctly and hope that the postal worker delivers it appropriately. Otherwise you'll be disappointed because your case remains unsolved.
>
> Port numbers are often a matter of a convention which becomes a standard. The longer a software or service has been in wide use, the more likely it is to have a dedicated and well-respected port number, such as 443 for https.

Nothing says you *must* make your Dogecoin node listen on port 22556, but if you don't, other people will have trouble connecting to you. Similarly, if multiple nodes for different coins all listened on the *same* port, the network would get clogged and congested and confused–so the convention of a different port for each coin is effective.

Chain Parameters

Using different network ports is a way to avoid collisions between the networks of different coins. What keeps a Dogecoin *address* from colliding with a Bitcoin or Litecoin address?

Part of the answer is "they all start from different places", but that's not a satisfying answer. All of these coins use the same ECDSA cryptographic mechanism (see Create Asymmetric Keys, pp. 17) and the underlying cryptography to make secret private keys corresponding to public keys which get hashed to make addresses. How is a Dogecoin address immediately recognizable as a Dogecoin address?

Bitcoin's source code uses a file called *src/chainparams.cpp* to define a few constant values used to produce and validate addresses. You can find the same file in Litecoin and Dogecoin. The values in this field govern how deterministic wallet derivation (see BIP-44, as used in Host a Treasure Hunt, pp. 174) works, how addresses are encoded, and how to validate all of this information.

BIPs are often Bitcoin-specific, even though many of them are useful for other coins. The design and code gets adopted where it makes sense. However, it's unlikely to expect that a Bitcoin proposal will go as far as to explain how another coin should work *even if* a few different parameters would make different things behave differently.

That's where SLIPs come in. The company SatoshiLabs, inventor of the Trezor hardware wallet, hosts a repository for SatoshiLabs Improvement Proposals, or SLIPs. These represent and often correspond to BIPs, but allow for multi-coin configuration. For example, SLIP-44[27] contains chain parameters for multiple coins.

A few parameters define how all of the world should interpret blockchain data, such that it's easy to distinguish between a Bitcoin value, a Litecoin value, a Dogecoin value, and any of the other myriad cryptocurrencies built around the Bitcoin design. In the Dogecoin source code, these parameters look like:

```
base58Prefixes[PUBKEY_ADDRESS] = std::vector<unsigned char>(1,30);
base58Prefixes[SCRIPT_ADDRESS] = std::vector<unsigned char>(1,22);
base58Prefixes[SECRET_KEY]     =
     std::vector<unsigned char>(1,158);
base58Prefixes[EXT_PUBLIC_KEY] =
     boost::assign::list_of(0x02)(0xfa)(0xca)(0xfd)...
base58Prefixes[EXT_SECRET_KEY] =
     boost::assign::list_of(0x02)(0xfa)(0xc3)(0x98)...
```

The emboldened values are most interesting; they're hexadecimal values corresponding to single text characters used as prefixes for addresses and keys. The first value, 30, will produce the character D when the address goes through the

---

[27] See https://github.com/satoshilabs/slips/blob/master/slip-0044.md.

Base-58 conversion[28]–the addresses that start with a A and are used for multi-sig and other advanced features. Dogecoin uses this mechanism to encode p2pkh addresses–the addresses you're used to dealing with. The second value, 22, will produce either A or 9 when the address goes through Base-58 encoding. Dogecoin uses as a prefix to identify p2sh addresses (see Put Funds in Escrow, pp. 206).

The third value, 158, is used as the prefix for WIF-encoded private keys (see Interchange Your Wallet Keys, pp. 15).

Finally, the last two values help encode extended public and private keys. Rather than single bytes, they are two, four-byte values. These values are used to manage derivation paths for HD wallets (see Derive More Addresses, pp. 318).

What Can You Do with This?

You don't have to know or even fully understand all of this information to work with Dogecoin effectively. However, once you start to see how all of the pieces fit together, you can also see how competing (or friendly cooperating) cryptocurrencies need to keep separate things separate. Even though Bitcoin, Litecoin, and Dogecoin use different mechanisms for miners to prove their work, they all use the same fundamental cryptographic structures. Similarly, they all use similar *implementation* with regard to network communication, blocks, transactions, script, and so on.

Furthermore, now you know more about the *mechanism* by which an address verifier can tell at a glance whether a long string of numbers or Base-58 digits is meaningful to the Dogecoin network or not–or at least, whether that string is definitely *not* meaningful for a given network.

  Tip #90   Measure Block Velocity

If you ran a poll of everyone with an interest in cryptocurrency "why is a long number representing funny Dog money worth anything?", you'd get some combination of three answers:

- "It's not. Is it?"

- "The price will go up in the future as more people want it in the future"

- "It's worth something because you can do things with it"

---

[28] See the Bitcoin wiki's list of address prefixes at https://en.bitcoin.it/wiki/List_of_address_prefixes.

If this polling is true and accurate, then you can divide the respondents into those who see a cryptocurrency as primarily an investment and those who see it as primarily an instrument of utility. This divergence in philosophy mirrors a similar debate in the investing world, where technical analysis measures patterns of price movement over time (predicting buyer and seller sentiment based on past price movements) and fundamental analysis measures the underlying value of a business, commodity, or service (trying to figure out what something is worth based on what it does).

This is complicated stuff, full of jargon, debates, charts, and theories. If you want to oversimplify, however, you can say that the fundamental point of fundamental analysis of Dogecoin revolves around the questions "what can I do with it" and "what can anyone do with it". To illustrate by analogy: if you issued your own dollars but no one else had any, they'd be worth only what you could talk someone else into giving you for them. That's probably not much. However, if you had 1 of 100 CoolBux and everyone in your friend group swapped CoolBux for pies or babysitting or pictures of cute dogs, then CoolBux is worth something because you can do so much with it.

Of course, even if 100 people each hold one CoolBux but never actually trade them, then their apparent value will go down because no one is actually using them.

Why the Economics Lesson?

This is nerdy stuff for a book about technology and Dog-themed money, but it's a risk to ignore economic fundamentals. Be skeptical of any simple explanation of human activity, but if this subject interests you, read about the Capitol Hill Babysitting Co-Op[a], which is one actual, studied example of a local currency as well as a Bitcoin-themed criticism of a common CHBC analysis[b]. Then read about "wage rigidity".

[a]See https://en.wikipedia.org/wiki/Capitol_Hill_Babysitting_Co-op.

[b]See https://nakamotoinstitute.org/mempool/babysitting-bitcoin-skeptics/.

If the model where usage influences value is correct, then one way to value cryptocurrency is measuring how much of it gets used regularly. We can do that.

Velocity and Velocity

The economic measurement of how much money circulates how often is called "velocity". If your author earns $1 CAD selling maple-flavored dried apple chips and buys $1 CAD worth of gasoline for a chainsaw and the chainsaw seller buys $1 CAD worth of poutine all in the same week, that's $3 CAD worth of velocity from the same Canadian dollar coin trading hands. How cool to watch a loonie spread through the economy: $3 CAD worth of value created from a single $1 CAD coin!

Similarly, if there were only 1000 Dogecoin in circulation, but all of them circulated between different people in every block, that's 1000 worth of Dogecoin value expressed in every block every minute or so. That's a lot of value created from a small amount of Dogecoin!

How do you actually measure that?

Counting the Dogecoin Circulating in a Block

At any specific block, some Dogecoin change hands, even if that's only the coinbase reward reaching one or more miners. How many coins circulate in a block? Calculate this by loading a block, looping through each transaction, and summing the outputs.

In pseudocode (see Decode Transactions, pp. 170):

```
def count_dogecoin_in_block(blockhash):
  block := RPCcall 'getblock', blockhash

  total := 0

  for tx in block.tx:
      raw_tx = RPCcall 'getrawtransaction', tx.hex
      tx_body = RPCcall 'decoderawtransaction', raw_tx.result
      total += sum(tx_body.vout.value)

  return total
```

For block 4,906,660, the result is about 380,000 Dogecoin circulated.

Counting All the Dogecoin in Circulation at a Block

There are two ways to count the number of Dogecoin in circulation at any given block. First, count all of the Dogecoin awarded as mining rewards for the coinbase transaction in every block (see Calculate Block Fill, pp. 255 and Analyze Block Metadata, pp. 270 for help). Alternately, you could estimate using the block reward schedule.

Find that schedule in *doc/FAQ.md* in the Dogecoin Core source code repository. For the first 99,999 blocks, the maximum mining reward was 1 million Dogecoin. From block 100,000 until 144,999 the maximum mining reward was 5 million Dogecoin. From block 145,000 until block 199,999 the reward was fixed at 250,000 Dogecoin. For the next several groups of 100,000 blocks, the reward halved at every threshold, until the reward stabilized at the current level of 10,000 Dogecoin per block since block 600,000.

The first couple of threshold had *random* rewards, where the reward itself was anywhere from 0 to the threshold. In other words, the reward for block 13 could have been 0 Dogecoin or 1 million Dogecoin or anything in between[29].

Given this information, it's possible to write an algorithm that estimates the maximum block reward at any point. For example, here's an algorithm in Perl including a test case[30]:

```perl
#!/usr/bin/env perl

use Modern::Perl '2024';
use Test2::V0;

test_rewards();
done_testing;

sub mined($h) {
    return $h * 1_000_000 if $h < 100_000;

    return ($h - 99_999) * 500_000 + mined( 99_999 )
        if $h < 145_000;

    return ($h - 144_999) * 250_000 + mined( 144_999 )
        if $h < 200_000;

    return ($h - 199_999) * 125_000 + mined( 199_999 )
        if $h < 300_000;

    return ($h - 299_999) * 67_500 + mined( 299_999 )
        if $h < 400_000;

    return ($h - 399_999) * 31_250 + mined( 399_999 )
        if $h < 500_000;
```

---

[29] In fact, it was 909,605 Dogecoin.

[30] The algorithm in a lazy, recursive, functional language is more interesting but it might obscure the logic behind cleverness. On the other hand, Haskell's QuickCheck is designed for exhaustive testing in situations like this!

```
    return ($h - 499_999) * 15_625 + mined( 499_999 )
        if $h < 600_000;

    return ($h - 599_999) * 10_000 + mined( 599_999 );
}

sub test_rewards {
    is mined( 1 ), 1_000_000;
    is mined( 2 ), 2_000_000;
    is mined( 99_999 ), (1_000_000 * 99_999);

    is mined(100_000), mined(99_999) + 500_000;
    is mined(145_000),
        mined(99_999) + (500_000 * 45_000) + 250_000;

    is mined(200_000), mined(199_999) + 125_000;
    is mined(299_999), mined(199_999) + 125_000 * 100_000;

    is mined(300_000), mined(299_999) + 67_500;
    is mined(399_999), mined(299_999) + 67_500 * 100_000;

    is mined(400_000), mined(399_999) + 31_250;
    is mined(499_999), mined(399_999) + 31_250 * 100_000;

    is mined(500_000), mined(499_999) + 15_625;
    is mined(599_999), mined(499_999) + 15_625 * 100_000;

    is mined(600_000), mined(599_999) + 10_000;
    is mined(699_999), mined(599_999) + 10_000 * 100_000;
}
```

Even if you don't read Perl[31] the logic should be largely straightforward. The only new idea you might encounter if you haven't programmed a lot before is the idea of *recursion*, where the calculation uses previous calculations of smaller values to winnow down the requested result to a single, final number. This is probably what you'd do if you were calculating this number by hand, however: calculate blocks of blocks, sum the rewards for each of those, then add the results together.

Given a block of 4,906,660 (freshly mined during the writing of this tip), the maximum number of Dogecoin in circulation is 203,253,110,000–or 203 billion.

Velocity Per Block

Divide one number by the other. With 380,000 Dogecoin transacted in block 4,906,660 out of a possible maximum of 203 billion, the velocity of this block is 0.000187%.

---

[31] It's not for everyone!

Is that good or bad? That depends. It's a number. Some blocks will be much higher (especially older blocks). Other blocks will be lower.

What Can You Do With This?

Economists don't measure velocity at a single, isolated point in time. They measure it over a period of weeks, months, and years. Economists probably *do* measure with rounder numbers than 0.000187%, but remember that the final calculation used an *estimate* of the maximum total Dogecoin in circulation as well as rounded numbers for the number of Dogecoin in the transaction.

To make this number more useful, calculate how many coins moved in a day or week or month and compare that to the total number of coins in circulation. You can also compare the velocity of Dogecoin–with its ever-increasing coin supply– to that of a coin with a fixed maximum number of coins. (This is, of course, most interesting for coins that have already reached the maximum number circulating.)

Remember as well that the blockchain uses a public ledger where all transaction outputs must be completely spent at every transaction. If you broke a $100 bill to pay for $1.50 worth of chewing gum at a convenience store, the blockchain would measure that as $100 moving around. These numbers don't work the same way way that non-blockchain currency numbers work. That's neither good nor bad–just differently interesting.

Can you increase or decrease velocity without adding value? Of course! Locking coins in a wallet decreases velocity. Losing coins sent to an address by forgetting your key decreases velocity. Moving coins around increases velocity. Moving coins between wallet addresses increases apparent velocity without necessarily increasing utility. We bring our own meaning to the numbers. They're just numbers.

There's no hard and fast rule to measure how useful a coin or a currency is other than, as an individual, how useful it is to you. Yet there are ways to calculate interesting values if you take the time to do so. With a public ledger and a little bit of math and programming, you have a wide range of information available. Do with it something useful!

  # Tip #91   Tip a Geocache

Back in the day, ages ago, to get where you wanted to go, you either had to have someone give you directions (people did this over the phone, and phones had cords connected to the wall) or plan your route with a paper map. Sure, a few hundred GPS satellites launched by the end of 1978, but you had to be really into satellites and coordinates to use them.

Then hand-held receivers appeared and became popular. Then phones became smaller and smarter and more capable. You may now be reading about Dogecoin on a device that can know exactly where in the world you are. Now you can get directions from almost anywhere in a populated area to almost anywhere else in a populated area (and many unpopulated areas too) with a few taps on a phone.

New technology adoption happens that way. Things start slow, with gradual adoption of weird or scary or complicated things that start to seem normal when cool and fun and useful applications arrive. That happened with GPS. It could happen with Dogecoin as well.

What is Geocaching?

Geocaching is a kind of collective scavenger hunt where people hide containers ("caches") and then post clues and coordinates to help others find them. Caches can be in busy urban areas (like a park) or in the middle of nowhere (like a forest). They can be small (an empty pill bottle or film canister) or large (a multi-liter bucket).

Visit https://www.geocaching.com/play to learn more about geocaching, including a free app for your phone to track clues and log your progress.

Larger caches often contain a log book, where people sign their names to record that they deciphered the clues. Many of these larger caches also include small take-and-replace trinkets, for a take one, leave one approach. These trinkets can be stickers, tiny plastic toys, or even low-denomination coins from far-off countries. This suggests an opportunity.

What is a Paper Wallet?

Most of this book assumes you're going to keep your Dogecoin in some kind of electronic wallet, such as the Dogecoin Core or a hardware wallet. Many tips also repeat the fact that a wallet is only one way to keep track of private keys and addresses. If you know the addresses that hold your Dogecoin *and* you have a way to represent those keys, you can access your funds even if your private keys aren't accessible through any hardware or software wallet.

If you lose your keys *everywhere* you lose your Dogecoin.

A paper wallet is a piece of paper containing data such as a Dogecoin address, private key, and QR code representing that key; it's a way to share that private key and address with someone else so they can claim the unspent funds for that address. It can be an effective way to give someone their first Dogecoin–such as tucking a piece of paper into a birthday card for your niece on her thirteenth birthday and proving you're the coolest aunt/uncle/relative ever.

To print a paper wallet with the Dogecoin Core, launch the GUI, then go to the File menu and select Print Paper Wallets. This will pop up a window showing what you're about to print: a single page image with a happy Dogecoin logo, a QR code, a line where you can write the number of Doge you sent to the address, and the private key/address.

When you print this page, you'll receive a prompt to send Dogecoin to the address to *fund* the wallet. To give your niece a cool 1000 Birthday Dogecoin, you must transfer that amount to the address shown on the paper wallet. Follow the prompts.

Note: your Core will not store the private key of this wallet. The Core generates a new private key and immediately discards it after you print the wallet. Any funds you send there will need the private key, so if you lose the paper, you lose the funds. If someone gets the paper but loses it or doesn't know what it is or destroys it, the funds are gone too. For anyone to redeem the funds, they must add the private key to a wallet that can access the funds immediately.

Put a Paper Wallet in a Cache

How do you tie these two ideas together? Print a paper wallet and fund it with a nominal amount of coins. Most trinkets you find in caches are worth less than a dollar, so (at current prices of about $0.10 USD per Dogecoin), 5 or 10 Dogecoin is a nice start.

Write out some basic instructions (like "Scan the QR code to claim your Dogecoin!" and "Take and scan this paper immediately!" or even "Hey, look at this cool website/FAQ/book for more information!") and put the paper in a sealed plastic baggie to protect it from the elements.

Now you have two choices. Either make your own cache or find an existing cache and leave the paper as a trinket for someone else to find. The nice part about making a new cache is the fun of making a puzzle (see Host a Treasure Hunt, pp. 174 for a variant idea you could mash up with this), but the nice part about randomly leaving a paper wallet as a trinket here and there is that the point *isn't* for people to find cryptocurrency, it's for people to *stumble across Dogecoin serendipitously*, discover its value, and find their own utility in the coin and community.

Understand the Risks

Depending on the geocache, the location, and the weather, water or dirt or mold or curious critters could get ruin the paper wallet. Hopefully someone finds the cache first, but there's always the risk that the coins could be lost forever.

Depending on how often and where you leave tips, you could inadvertently reveal your personal information (name, approximate location, regular travel routes, et cetera) and that of anyone who redeems one of your wallets. If someone recognizes what the paper wallet represents, leaves it in place, and signs their moniker to the log, they can return later and see other log entries and narrow down who might have taken the wallet. If they scanned the paper, they have the keys to the wallet and can do anything with the transactions.

Even if they didn't scan the paper, they can now tie a list of names to a wallet address and monitor all transactions to and from that address.

It's up to the person who takes the wallet *paper* to practice good transaction security by immediately transferring the funds to a new and secure wallet address. As with any paper wallet you weren't immediately and directly handed by someone you'd trust with your wallet address, treat the wallet as potentially hostile.

Finally, some people might see the chance to add a few more Dogecoin to their holdings as a personal challenge and try to find all the wallets you leave. That might go against your purpose in randomly tipping strangers who might have never encountered Dogecoin before. Consider what you want to achieve–hopefully brightening someone's day and inviting them to try something new, enjoyable, and liberating–and if your actions are likely to achieve that.

Tip #92 Build with Dogecoin

Some people want to enhance or replace the global financial system. Some people want to send money across the world for fractions of a cent in seconds. Some people want to make micropayments and tiny transactions feasible for an Internet-connected world.

If the only thing we ever did with our funny, dog-themed cryptocurrency were to trade memes, have fun, and tip each other back and forth that would be enough. We can do more.

All of these are possible. All of these are happening. Some of these are the precursors to other new, interesting, useful, creative, and fun things. For example, this book includes a chapter on using Dogecoin as the payment mechanism for coin-operated systems such as pinball machines, jukeboxes, and vending machines (see Manage a Dogecoin Arcade, pp. 301).

Putting the pieces together for *your* project is up to you. Building that system well is a lot easier if you're working with a community of smart, fun, helpful people. That's one of the goals of Dogecoin: do good things together?

Find Your Project

Working with other people to build something amazing sounds amazing, but without something specific to build, you might feel like the business school graduate going to the computer science job fair and trying to find a cofounder for your great new startup idea[32].

What motivates *you*? What problems do you face? What doesn't exist, but if it did, would make your life easier or more pleasant or more enjoyable? What frustration would you like to go away?

When you close your eyes, in the moment between the click of the light and the start of your dream, what idea won't stop dancing in your brain? *That's* your project. When its shape starts to become real, grab something to take notes about what it really is. What does it do? What doesn't it do? How does it feel to use it?

Write down your ideas. Draw pictures. Make lists. Write a press release from the future as you're bragging about shipping the first version of your project that you can be proud of.

Then make a bunch of little lists of how the features start to work together. This won't solve all of your problems, and it won't magically make things appear, but it will help you talk to the right people about the right things–especially if you know the two or three essential things you absolutely have to have. Then you're ready to start building with other people.

Find Your Community

You may think "But I'm not a developer" or "I'm only a junior developer". What can you do? Plenty! For example, although your author wrote a lot of code and proofs of concepts and so far has managed not to destroy any hardware proving that his ideas will work, he's only one person with some ideas, some implementation, and the stubbornness to manage a lot of things he understands by himself.

If you were to adopt his code for your own purposes, you might find that it doesn't work in some obvious ways, or that it's missing important features, or that you're missing some essential customization. All of these can be the basis of a useful *conversation*–ideally one before you've struggled with someone else's software and cried out in frustration.

Your requirements–the essential things you absolutely must have–are vital to the conversation, especially for good developers who want to build things that other people find useful and usable.

---

[32]Especially if your startup idea is "Let's make a really popular app!"

You might find yourself challenged on what you want to build and why. A good, healthy community makes this an act of creativity and collaboration. "Why do you want to do *that*?" can be a question motivated by genuine curiosity and care, especially for novel and unique ideas.

This is where the cooperation comes in. You may have to explain your idea and you may have to explore other implementation options, even going so far as trying software under development with an open mind.

Create or Fork Your Community

If you can't find a community that's working on the same things you are, you may have to build one. If you can't find a community where you fit and can work, you may have to fork it. That's okay too, and it's not as violent as it sounds. Think of a road which splits into two diverging paths to go around a hill. The paths still end up going somewhere useful and they both acknowledge the obstacle, but they take different approaches.

The world of free and open source software practices this on a regular basis. The license of the code itself often supports this behavior[33]. That's where Dogecoin came from—a fork of a fork of Bitcoin source code, with unique changes over the years and other changes borrowed liberally and thankfully from other projects. We work better when we work together.

You're also always welcome to start your own project from an empty directory and a good idea. That's how many projects start too!

Join an Existing Community

If you want to join an existing community to find like-minded people, consider working with the Internet of Doge[34] community. This is an incubator for Dogecoin-related projects including:

- The aforementioned arcade machine payment system

- A sponsorship program for creators, like Patreon, but paid in Dogecoin

- A Dogecoin-based crowdfunding platform

Another exciting community is very.engineer[35]. This project focuses on building

---

[33]But make sure of this before you start over with existing code!

[34]See https://github.com/chromatic/internet_of_doge.

[35]See https://very.engineer/.

hardware systems for Dogecoin, including small, inexpensive devices to run Core nodes.

Finally, the Dogecoin Foundation has several projects built on and around the Dogecoin network protocol. Check the Forums[36] to discuss these and like-minded projects:

- A payment processing system for merchants (invoice-based)

- An embeddable shared library to work with the Dogecoin network protocol and addresses

- A project to relay transactions over long distances using radio signals

Understand the Risks

Working with other people can be as difficult as it is rewarding. Sometimes people get hung up on their own ideas and ways of working. Other times people are quick to criticize and slow to help. Sometimes people are mean.

You can control only yourself. If you're working with other people, set your boundaries and expectations, and influence the community to grow in healthy ways and avoid unhealthy behaviors. This doesn't mean avoiding disagreement or conflict. Those can be good if handed well–discussions can lead to better ideas and clearer understandings, especially if you're willing to speak (and, sometimes, change) your minds.

As a concrete example, your author does most of his prototyping work in Perl, because he's used it for a long time and knows a lot of the tips and tricks[37]. However, he's comfortable working in other languages and willing to learn new ones. If writing the backend services for a Dogecoin project in Python means that more people can contribute, he'll do that.

There may be *technical* reasons to prefer Perl or Ruby or Golang or Rust or Node.js or PHP or any of a hundred other languages, but the human factor of people's contributions is even more important. The problem's we're *trying* to solve with programmable money is how to work *together* without giving up our own power and without centralizing authority in hands we can't trust.

---

[36] See https://forum.dogecoin.org/.

[37] See his book *Perl Hacks*, for example, at https://amzn.to/3RfIho6.

## CHAPTER 10

Manage a Dogecoin Arcade

As if this book weren't nerdy and technical enough, here's a full, working example to tie everything together: a design and implementation for an arcade taking payments in Dogecoin.

This example has a lot to recommend it. It brings Dogecoin into the real world. It represents small, self-contained transactions. It uses the network for its strengths: making small payments quickly.

Regardless of whether you have actual video game hardware, emulation or virtualization systems, or no hardware[1], this system's design and implementation will work for many types of payments made or received on the Dogecoin network. You might run a coffee shop or volunteer at a farmer's market or have a great bake sale fundraising idea or any of a thousand other things.

You might have physical hardware too: a pinball machine by your front door, or a claw machine, or a cold drink dispenser or something even cooler we haven't invented yet.

Bringing Dogecoin into the real world requires some integration, but all of the parts are within your reach. Let's start by assuming you have a combination of hardware and software.

  Tip #93   Associate Addresses to Machines

Suppose you have a small retro arcade[2] in the back of your family's ice cream shop. You run a little network to handle the display screen showing your ice cream

---

[1]Maybe you have little interest in arcade games. That's fine too!

[2]Again, this could be a single machine, drink dispenser, or pachinko machine: anything coin-operated!

flavors and for the payment/inventory system you use to sell waffle cones, scoops, and sprinkles.

With a multi-machine arcade, if you want to take Dogecoin in exchange for credits in your arcade machines, you need to know which machine has received a specific customer payment. There's an easy way to do this and a difficult way to do this.

An Address Per Machine

The difficult way to do this is to run some kind of node on each machine that knows one or more addresses. What makes this tricky?

- you need node-capable hardware for each machine

- you need storage space, electricity, and bandwidth for each machine

- you have maintenance costs for each node in each machine

The benefit of this technique is that you've successfully associated an address (or multiple addresses) with the machine when you configured them. If you want to swap your Ms. Pac-Man console for your buddy's Galaga console, you can change the hardware and everything's good.

If, on the other hand, you have a single node that can monitor the network for transactions, you can watch for transactions for *all* of your machines and route requests to individual machines. You need to design and configure more software in this case, but you get more flexibility.

Tracking Addresses

If you've configured each machine with some sort of network connection (see Program Real Buttons, pp. 321), you might have machine names mapped to IP addresses from your DHCP server[3]. Follow a few steps to get up and running.

Generate a Bunch of Addresses

First, you need a bunch of addresses, maybe exported from a trusted node (see Export and Extract Wallet Data, pp. 153) or generated from a secure HD key. Prepare these in a text file to import into a database.

---

[3]If this seems more technical than you want to manage right now, trade a couple of pints of ice cream with an IT-savvy friend to set this up!

Configure Your Network

Suppose you have five arcade machines. You also have five DNS entries for them:

- `pacman.tastytreats.local`

- `centipede.tastytreats.local`

- `crystalcastles.tastytreats.local`

- `indianajones.tastytreats.local`

- `babypacman.tastytreats.local`

Assume that each of these machines has *something* reachable on the network. Now you must send some kind of action or event to these machines when an associated address receives an appropriate payment. How do you do this?

You *could* use a JSON file stored in a directory somewhere (see Enhance RPC Calls, pp. 80 for an example). That's easy for a human to update and easy to read. It's less manageable if you want to access it from multiple places simultaneously or if you want a nicer interface than JSON in a text file.

Assign Address to Address

An alternate approach is to create a database, such as with SQLite (see Export Block Metadata, pp. 263 for an example). In a larger environment, you might prefer MySQL or PostgreSQL to handle multi-user capabilities, but that adds management complexity you might not be ready for yet. A table to map Dogecoin addresses to machine names may resemble:

```
CREATE TABLE addresses_to_machines (
  address CHARACTER(64) NOT NULL,
  dns_name TEXT NOT NULL,
  is_active BOOLEAN NOT NULL DEFAULT false
);

CREATE INDEX addresses_to_machines_address_idx
  ON addresses_to_machines(address);
```

These statements tell SQLite how to represent mappings between Dogecoin addresses and machine hostnames on your network. A series of INSERT statements (or a CSV import) could populate this table. You can assign addresses to hosts randomly, let your software do it (see other tips), or use any alternate approach.

Update Addresses

The `is_active` field allows you to toggle an association on or off. Depending on how you route information to each machine, you could burn an address and take a machine off the network entirely (someone dripped rocky road on the Crystal Castles trackball, so there's no point in adding plays to the machine, for example).

Note also that there's nothing that says each machine can have one and only one DNS address (give your IT consultant an extra pint to pull this off **or** to discourage them from pulling this off) and there's definitely nothing preventing you from associating multiple Dogecoin addresses with a single machine. In fact, that might be a good idea (see Rotate Machine Addresses, pp. 307).

To add more addresses, import or add them to this table.

What Can You Do With This?

What happens now? You have enough to focus on the remaining tasks:

- tracking transactions as they come in

- turning incoming transactions into interesting events

- handling interesting events on your arcade machines

This abstraction described here is straightforward, and it adds flexibility and power to your system. You can deal without it if you have smaller or simpler needs, but you need something like it at some point as your arcade grows.

Understand the Risks

Hardware, network, and physical security is always a risk. If you use a wifi network, make sure that you have an isolated network with the appropriate password and security settings to keep other people from connecting to your network, lest they have an easier time to connect to *any* of your machines.

If you have a wired network, wifi security is less of an issue, but you may have to run a lot of cable and deal with more physical security, lest someone plug their own hardware into an open network port.

Hard-coding an association between address and machine makes certain things easier (see Customize Your QR Code, pp. 149), but it reduces your anonymity. Any visitor to your retro gaming museum who pays for individual plays against specific machines knows the Dogecoin addresses under your control and can monitor transactions against them and tie your physical location to your arcade income.

On the opposite side, if you keep logs of high scores, attendance, and the like, you can tie the addresses of your customers to their identities as well. While some people may be comfortable with their wallet security and pseudonymity, you should limit data retention and make any identification opt-in only with informed consent.

  # Tip #94   Practice Safe Wallet Hygiene

You stand in line at a bakery. You prepare to defeat a Balrog in a pinball arcade. You grab your favorite band's t-shirt and head to a concert. You have your phone loaded up with enough Dogecoin to get a tasty treat, smack some pinballs, or move to the music.

Would you pull out a wallet with a million dollars in it to spend $5 on a pastry and a beverage? Or break a $1000 bill to play four games of pinball? Or swipe a credit card with a $50,000 limit in a reader held by a teenager in an orange vest standing outside a stadium?

Hopefully the answer is no. Hopefully the answer is still no if you change "dollars" to "Dogecoin", because there's an even more important difference between physical currency and cryptocurrency: your cryptocurrency transactions are *public*.

Understand the Risks

Suppose you have 50,080 Doge in your wallet associated with a single address, and the doughnut shop has a QR code on the counter. "Scan it with your phone and hit send", the clerk says, noticing the Dogecoin pin on your lapel. The tape's coming up around the edges of the QR code. It's been there for a while.

The guy behind you in line stares as you send your transaction. He should be eyeing the doughnuts with that much interest.

You grab your buttermilk blueberry bars and walk outside. A moment later, your phone dings. Your balance is now 50,000 Doge. That's when the sketchy guy bumps into you and palms your phone, leaving you with doughnuts and crumbs. Hey–he stole your phone and your Dogecoin!

Multi-Use Addresses Are Bidirectional Risks

What went wrong? On your side, you kept all of your eggs in one basket. Wallet addresses are free, and[4] cryptographically secure. It's so unlikely you'll acciden-

---

[4]To the best of our knowledge in 2024!

tally bump into someone else's wallet address that you might as well guess the coordinates of a specific atom in the entire universe from the big bang[5]

Every time you reuse a wallet address to *send* funds, you run the risk of tying your identity to a specific address as well all of the addresses used to get funds out of a coinbase to that address. If you're spending Dogecoin in person, this risk is higher; you've potentially tied that chain of coin custody to your physical presence.

Multi-Use Addresses Can Be Snooped

What else went wrong? Anyone who knows that your local doughnut shop has re-used its address can track all transactions *to* that address. The sketchy man behind you in line could have looked up the shop's address on a blockchain explorer, found your transaction (the one that just happened a moment ago), then looked up the unspent transactions of the wallet address you used to pay for your tasty treat.

There's no excuse for any sketchy person to swipe your phone. They own their bad behavior. Even still, the ability to do bad things is multiplied by the shop's poor address hygiene.

Fortunately, this can all be fixed.

Limit Your Walking Around Money

Unlike a physical wallet full of $1000 bills, your wallet addresses can only be unlocked with your private keys. If they're not on your device or in your head, someone will have to do a lot more work to get them.

Instead of keeping all of your funds available on instant access by swiping your finger on your phone, store the bulk of your funds in a more secure location: an offline wallet, a hardware wallet, or backed up with keys not connected to the Internet or available by any easy network access.

A few hundred Doge here and there in various addresses spreads out your exposure. Losing access to (deliberately or by accident) one address is still unpleasant, but it's not the catastrophe of losing all your funds.

Encourage Vendor Address Hygiene

When you're out and about spending your favorite friendly dog-themed money, pay attention to vendor practices as well. If you see someone re-using addresses for purchases, consider taking the time to inform them (gently and kindly) about the risks of re-using addresses (especially for their customers) and benefits of

---

[5]This was a plot point in an X-Men comic to explain a mutant's overpowered prescience.

address isolation (especially for their customers). Buy them a copy of this book, if necessary.

With the prevalence of HD wallets (see Use a HD Wallet, pp. 39) and other mechanisms to generate multiple addresses managed by a single key, there's no reason not to take advantage of the nearly-limitless address space of the blockchain.

Maybe it takes brainstorming, like "Hey, what if you could print a QR code on the initial receipt?" Other times you might have to get together some software-minded folks to figure out the right approach, such as generating the next arcade game's QR code and displaying it in the cabinet's attract mode (see other tips).

Security is often a balance between convenience and inconvenience. The question is how far you want to go to protect yourself, your vendors, and your customers and how much extra effort to expend to extend that protection. The core developers and advocates can only go so far giving you the tools and the information you can use to do what you want to do easily and safely. The rest is up to you.

Tip #95   Rotate Machine Addresses

Your little arcade is coming together and you're excited! You've set up a way to track addresses and machines, and you can reach every machine over a secure network with its own name (see Associate Addresses to Machines, pp. 301).

It's mid June[6] and the ice cream business is booming. It's time to get people enthusiastic about playing games before, during, and after enjoying a tasty frozen treat. You haven't yet printed QR codes to tape to each machine, because you're still thinking about how to handle customer data safely (see Practice Safe Wallet Hygiene, pp. 305).

What do you do?

Back to the Database

The `addresses_to_machines` SQL table had an interesting column, `is_active`. It implied that a machine could have multiple Dogecoin addresses. How would that happen?

Suppose you have a table full of addresses under your control:

```
CREATE TABLE wallet_addresses (
  address CHARACTER(64) NOT NULL,
```

---

[6] . . . as of the time of this writing!

307

```
  label TEXT NOT NULL DEFAULT ''
);
```

Find Unused Addresses

Maybe you've generated these with from an HD key and derivation paths or you've exported them from a wallet (see Export and Extract Wallet Data, pp. 153). When they're in your database, you can do several things. First, to see which addresses you've already used:

```
SELECT address, label
FROM wallet_addresses
WHERE address NOT IN (
  SELECT address
  FROM addresses_to_machines
);
```

This query looks for all addresses associated to machines, then looks up all addresses in `wallet_addresses` that aren't in that list[7]. That second SELECT nested in the outer SELECT is called a sub-select. The database will look up all the results for the inner query, then use its results to filter the outer query.

If you run this query repeatedly, turn it into a view:

```
CREATE VIEW IF NOT EXISTS unassigned_wallet_addresses
AS
SELECT address, label
FROM wallet_addresses
WHERE address NOT IN (
  SELECT address
  FROM addresses_to_machines
);
```

Add New Machine Addresses

Now you have a set of unused addresses to go along with the set of used addresses and the set of active addresses and the set of machines. It's important to think in sets of data when working with a database. What's the important data flow here?

For every machine, disable the currently active address. For every machine, get a new address. For every new address, mark it active. You might write SQL like:

---

[7]Experienced database users can find better ways to write this query, so if you find this kind of work interesting, try some yourself as a learning opportunity!

```
BEGIN;

UPDATE addresses_to_machines
SET is_active = FALSE;

WITH available_addresses AS (
  SELECT
        ROW_NUMBER() OVER (ORDER BY address) row_number
      , address
  FROM unassigned_wallet_addresses
  ORDER BY RANDOM()
),
available_machines AS (
  SELECT
        ROW_NUMBER() OVER (ORDER BY dns_name) row_number
      , dns_name
  FROM addresses_to_machines
)
INSERT INTO addresses_to_machines (address, dns_name, is_active)
SELECT address, dns_name, true
FROM available_addresses
JOIN available_machines USING (row_number);

COMMIT;
```

There's a lot going on here! The BEGIN and COMMIT commands tell the database to treat the entire set of statements as a single unit that either succeeds entirely or fails. If there are any errors, the database will discard *all* of the changes. This prevents you from getting your data in an inconsistent state due to a logic error. It won't prevent you from writing the wrong code, but it will prevent any execution failures from persisting.

The first UPDATE statement disables all currently associated addresses in preparation for the new addresses. You might not want this behavior; it's safe to remove if you like. However, it does show the value of the transactional behavior of BEGIN and COMMIT. If everything goes correctly, the new addresses will now be active. If that fails, the old addresses will still be active.

With the transactional behavior, either success or failure means that the machines will keep running. Without the transactional behavior, this code could end up with *no* addresses active. Oops!

The next statement looks like three statements, and it kind of is. The WITH ... AS behavior looks like the sub-select approached used earlier. It's a technique called Common Table Expressions, or CTEs. Naming part of the query this way makes it available as a data source in other parts of the query.

We want to use a list of machine names and a list of addresses as a single data source to insert new records into the list of machine address associations. This query thus has two CTEs.

One remaining piece of the query is tricky. What's the relationship between data in the address table and in the machine table?

From the view defined earlier, you know that *some* rows in the addresses table *may* be in the associations table, but ideally most address rows *aren't* in the machines table. Because the first CTE selects from the unassigned address view, we know that there is *no* association between rows in each table.

This is a problem.

SQL thinks of tables as sets *and* it allows you to join sets to each other *if* they have common characteristics. For example, if your ice cream arcade takes off and you open multiple locations, you might add a location identifier to each machine in the associations table and you probably will add a locations table to keep track of location-specific information. If you have a location column in each table, you can join those tables together to make an interesting query (show me all of the machines in our downtown food cart popup location, show me all of the machines grouped by their location, etc).

Given that there's *no* such association between available addresses and machines here, we have to make one.

The strange looking ROW_NUMBER() OVER (...) syntax here tells the database "arrange the results from this query in a specific order, then give each one a number starting from 1".

Available addresses come out in random order (that's what RANDOM() does). Machine DNS names come out in alphabetical order. All of this complexity allows the piece of SQL JOIN ... USING (row_number) to work. Where there was previously no relationship between these two sets of data, this code creates an artificial relationship and then uses it.

The INSERT INTO commands tell the database the name of the table, the columns of the table, and how each row should look. SELECT ... FROM is a great way to put data from one set (the union of the two CTEs) into the table. Note that each column comes from a different source: address from the randomized list of available addresses, dns_name from the list of machine names, and true as a hard-coded value.

Do It Now!

Run this code whenever you want to rotate the available addresses for your machines. Then fetch the new active address and machine name from the database and print up new QR codes, ship the information to each machine to display new

QR codes, or update anything else.

Understand the Risks

This isn't perfect, but it's better than using a single wallet address for all of your machines or using a single wallet address for each machine.

You can automate the address rotation on a fixed schedule, but if you have printed QR codes, someone will have to walk to each machine and update them. Maybe that's not a problem, but consider the human factor in the equation. Someday, someone will trip over a loose shoelace and send a stack of index cards flying, so that a customer who wants to play Super Street Fighter 2 Deluxe but gets a credit for Super Street Fighter 2 Championship Edition instead will be super complex disappointed.

If you have a way to rotate addresses on a schedule instead *and* you remove the need for humans to update QR codes or payment information or whatever tells a customer which address to use, you can make this system more secure and less fragile. Perhaps that's adding a QR code to the attract display of each machine, updating a webapp with the new address, pushing a new QR code to an LCD display on each machine, or something else.

  # Tip #96   Manage Tokens

Step back from the details of running a Dogecoin-powered arcade to consider what you want to accomplish. Sure, it's a clever hack to have an address for every machine and to be able to kick off a play session for every *n* Dogecoin received for the address associated with a machine.

It's worth *trying* to accomplish that. However, that technical approach has draw-backs.

First, transaction speed is an issue. Waiting an average of 55 or 60 seconds for a transaction confirmation a long time for someone to stand in front of your Atari Star Wars machine waiting to fly an X-Wing into the Battle of Yavin. If blocks are full, it could take longer or cost your customer more to send that transaction. You *could* work around this by checking the mempool for a pending transaction, but that's additional work and the transaction might not clear soon or ever.

Second, you have to deal with the association between a person sending a trans-action to a machine address and the machine address being visible to other people in the same location. If core developer Patrick Lodder really wants to play *Donkey Kong Jr.* but worries that the person getting ice cream behind him will corner him in the parking lot and demand 1000 Dogecoin from Patrick's Big Whale Wallet, you've lost a good customer.

Third, you have to mange the overhead of associating machines with addresses and keep that infrastructure running. It's not a lot of overhead *if* you're comfortable managing infrastructure like this, but it's still overhead. You can bribe the nice neighborhood kids with fudge ripple only so many times before they realize they can build their own arcades.

None of these are dealbreakers. You can work around them all, and they might be the right choice. However, you have alternatives.

A Token Economy for Your Dogecade

If you hear the word "token" related to cryptography, you might think of a scammy influencer shilling a Ponzi scheme and hoping to make a few thousand dollars of Ethereum before pulling the rug out from underneath their unsuspecting victims. That's definitely one definition of token.

Here it means something else: a digital asset redeemable in your arcade. There's no secondary market. There's no airdrop. There's no pump and dump or price or listing on some exchange or website. It's just a token that someone can redeem for a few minutes of nostalgic pixelated fun.

How does this work?

Buy Tokens with Dogecoin, Redeem Tokens for Game Credits

The high-level software design is straightforward. A customers comes to your arcade. You hand them a Dogecoin address (a printed QR code, a website, maybe it's a QR code that takes them to your website). They transfer 100 Dogecoin for 100 tokens.

Every machine has a QR code on it (or your website lists your games). Your customer selects "play this game" and your software subtracts a token from the customer's account and adds a credit to the game.

Skip over the "add a credit to the game" piece for now (see Flip a Switch, pp. 325).

To make this work, you need a steady and unique supply of Dogecoin addresses (see Export and Extract Wallet Data, pp. 153). You need a way to give a unique address to a customer and watch for transactions to that address. You need to account for tokens purchased from that transaction. You need a way to associate a customer's identity with the tokens they've purchased. You need a way to track token *usage* or *redemption*.

You're going to need to build an accounting system and a user identity management system.

Advantages of this Model

This approach adds predictability to your arcade model:

- On-network transaction speed applies only to token purchase, not token redemption.

- The price per token can float with your local Dogecoin to fiat currency exchange rate (good for you and your customers).

- Advance token purchases let you realize income sooner than on-demand redemption.

- Advance token purchases let you offer bulk purchase discounts to your customers.

- The association of Dogecoin address to purchased token pool offers more identity privacy for both you and your customers.

- Token purchases don't have to take place in your physical location. (Neither do redemptions, but that's up to you.)

Disadvantages of this Model

This approach has some drawbacks:

- The cool factor (I sent Dogecoin to an address and the arcade machine in front of me just lit up) isn't there.

- You have to manage customer identity in your transaction system.

- Customers may not want to keep a big balance of tokens, so your incentives for purchase management may need careful alignment.

System Design

That's a lot of theory! What does the system look like? Several other tips describe various pieces of this system, including Dogecoin address receipt actions (see Act on Wallet Transactions, pp. 194), a general webhook architecture (see Post to Discord, pp. 64), a list of machines (see Associate Addresses to Machines, pp. 301), and the magical "insert credit" automation bridge (see Flip a Switch, pp. 325).

This approach needs two new systems and an architecture design that ties everything together.

The first new system lets customers create persistent accounts in your arcade so that they can buy arcade tokens. If you assume everyone in your arcade has a phone with a QR code reader or web browser, you can simplify the design. Users sign up with an authentication mechanism, such as an email address and password or even just a durable session cookie[8].

That identity management system needs to connect to the token management system. Buying and redeeming tokens means you need to hand someone a unique Dogecoin address to receive payments. The receipt action and webhook system will deposit tokens in to the customer's account upon receipt. The customer must have access to their tokens and the ability to select one or more token to play on a machine.

Maybe this is a QR code on the front of every machine with a link to your webapp. If the customer has authenticated *and* has tokens in their account, the app can prompt to redeem one or more tokens for game credits. Alternately, an unauthenticated customer can be prompted to log in or create and fund an account.

Another direction is to have your webapp list all machines (perhaps with a map of your facility, if you've expanded to take over a small converted airplane hangar) and let the customer select which machines get credits. Again, you have to ensure that the customer has authenticated and has enough tokens for redemption.

As your system expands, you'll also want administrative functions, such as checking token balances, resetting passwords, updating machines, et cetera. If you're not a software person, this may sound like a lot of effort. If you're a software person, this could be enjoyable to build.

Understand the Risks

If you use this design, you'll have to write or manage more software. It's not a *lot* more software, but it's additional infrastructure for you to adopt, use, and maintain. You may have to deal with user identity management ("I lost my tokens! I switched phones and can't log in!") as well as connectivity issues between your token accounting system and your other systems.

Introducing an identity management system reduces anonymity in your system. For example, the straightforward way to allow customers to log in with their email addresses is to store those addresses in plain text. This implies an association

---

[8]The drawback of the cookie is that it's tied to a specific browser and device and, if lost, will require customer support to re-associate a person with an account or give them a refund.

between an email address and a Dogecoin address, which anyone with access to your database can use to de-anonymize transactions to that email address.

A *safer*, more private approach is to use a cryptographic hash (see Make a Hash of Fingerprints, pp. 1) of a customer's email address. Every time a customer tries to log in, hash their email address and compare it to the stored hashes in your system. If you've found a match, you can proceed to validate the provided password. Be sure to use a secure hash, however, lest you leave yourself vulnerable to dictionary or precalculation attacks.

Bcrypt everywhere?

The bcrypt[a] function is a tempting approach because it resists some of the more effective attacks that you might otherwise make against a system, *but* it's more useful for password checks than email checks, because the best way to use bcrypt is to have a different random salt and iteration count for each hashed value. Given that checking an email address is asking "does this account exist at all?" while checking a password is "given the salt and iteration count, do I come up with the same value?" the questions are very different.

Limiting your system to a single salt and iteration value for email addresses reduces the security of your system.

---

[a]See https://en.wikipedia.org/wiki/Bcrypt.

Depending on your local jurisdiction, you may also be subject to all sorts of financial rules and regulations[9]. As with any business venture, seek professional advice about tax, legal, and accounting issues.

  # Tip #97   Sell Event Admission

The properties of the blockchain (the ledger is public, transactions are immutable, data is distributed) and the properties of wallet hygiene (your private keys are private, you don't re-use addresses, you can prove your transactions are yours) can combine in interesting ways.

---

[9]If you've never heard of *escheatment*, you're probably doing something very right in your life.

Not all of them require immediate gratification. Some take place over time.

Suppose you have friends who put on a holiday-season, barn-raising concert in a little farmhouse just outside of Cincinnati, Ohio. They have a band and a bunch of good songs. You want people far and wide to bring joy to the darkest nights of the year. They've asked you to sell tickets to the event. They have a flexible pricing model where people pay what they want, and you don't want to tie anyone's identity to the amount they paid.

How do you let people into the barn, knowing they've purchased tickets, without setting up an entire payment verification and validation infrastructure? Also you don't want to mail out paper tickets, for all of the reasons that paper tickets are awkward and weird[10].

Some of the same infrastructure you use to manage your arcade/ice cream parlor can work here.

Share a Semi-Secret

Even though the Dogecoin ledger and all its transactions are public, you don't have to reveal your own identity or the identity of anyone you transact with. Yet for event admission to work, you need some way to prove that you made a transaction.

One solution is to exchange some kind of semi-secret.

For example, you can start with the same kind of system you use to fund tokens in your arcade (see Manage Tokens, pp. 311). Someone comes to the concert's website, hits the "buy tickets" button, and gets a QR code. On the backend, you generate a new, unique receiving address (similar to Rotate Machine Addresses, pp. 307) and set up an event listener for transactions to that address (see Act on Wallet Transactions, pp. 194).

You have a fork in the road here. You *could* ask people to enter an email address to which you'll send proof of payment. This is a standard practice for events, but it trades privacy for expedience (customer support may be easier). Alternately, you could generate an admission QR code or secret passphrase *on the spot* before payment.

This might feel backwards. Why give someone a ticket before they've paid? It's not backwards though; the ticket doesn't grant someone admission. It's your *validation* of the ticket at admission time that grants admission–the ticket itself grants nothing.

Think of it this way: you're not going to re-use the address you generated for the

---

[10]On the other hand, you can't autograph a smart phone the way you can a ticket.

purchaser (for hopefully obvious reasons). It's just an entry in your database that hangs around until it's outlived its usefulness (after the event or when you delete unused records).

If, when you give this unique address to a customer, you also generate their unique admission code, you can store them together in the database. Then your event listener can wait for the appropriate transaction to that address and mark the admission code as valid.

At the night of the event, at the door of the barn, all your system has to do is ask "is this code valid". If the code doesn't exist, it's obviously invalid. If the code exists but the address hasn't received funds, the code is not valid. You can even detect that latter case and say "Your transaction didn't go through; please try again." and wait somewhere between 55 and 65 seconds.

Keep in mind as well that the person showing you the code on their phone (or a folded up piece of paper) doesn't have to be the person who made the transaction. They may have finished babysitting goats for their uncle and requested payment to the concert address–and that's okay.

Understand the Risks

No one gets into the barn without some kind of ticket or name on the guest list. The minimum bar of admission is a printed QR code or an image on a smart phone or something else. Anything physical can be lost, destroyed, damaged, or stolen. Yet the network survives. This is a benefit of a distributed ledger.

What if you lose your ticket? Can you still prove that you made that transaction? If you can–if you've retained the appropriate private key–then that level of proof might suffice.

Alternately, if you know the unique, used-only-once, unguessable address you sent the funds to, you can prove that you made the transaction. At least, you can mostly prove it; someone else could be monitoring your transactions or may have a copy of your wallet. The same goes for paper tickets, however, or a QR code you received in your email.

The biggest risk of this scenario is tying someone's identity to a specific transaction. A concert with hundreds of attendees, each using their wallets safely and securely, may be less of a risk–but if you have that one friend who is a little too cavalier with their coins, consider gently explaining that address re-use is as bad as slow music played loud is good.

Tip #98   Derive More Addresses

Whether you use a batch of Dogecoin addresses for your machine (see Associate Addresses to Machines, pp. 301), generate new addresses for each new customer (see Manage Tokens, pp. 311), or otherwise receive Dogecoin in transactions, you need a steady supply of new addresses. This implies a steady supply of new public/private key pairs.

Wallet derivation (see Use a HD Wallet, pp. 39) simplifies this problem by describing a hierarchy of keys. Given a single starting key, you can generate an infinite number of related keys as often as you need them. To spend transactions sent to those addresses, you need the starting key and information about derivation but you don't have to track all of the public/private key pairs along the way.

Improve your security and increase your convenience by reducing bookkeeping? Yes!

Derived Keys

Key derivation follows BIP-32 and BIP-44 standards. You must understand BIP-44's notion of derivation paths. These are descriptions of *how* to generate additional key pairs from an existing master key. A Dogecoin derivation path looks like m/44'/3'/0'/0/0. The first three path components are always the same for BIP-44 paths; they indicate a derivation path (m), the purpose of the path (44'), and the Dogecoin network (3'). Subsequent path components are up to you.

The fourth component represents a unique account number. If you have multiple arcade locations, you can use different numbers for each location or machine or whatever. The fifth component is either 0 or 1 and indicates whether you intend for the derived address to receive external payments. You probably want to use 0 here. Finally, the final component is an index number that increases with each new address you derive.

When a conforming wallet sees a key and a derivation path, it can follow that path to examine the blockchain to see which addresses have ever received a transaction. Even with no unspent inputs associated with that address, if the derived address has ever received funds, the wallet can determine this—and know to use the *next* unused index value to derive further addresses.

You don't *need* a wallet to derive further addresses, though—only software that understands this derivation scheme.

318

The documentation around BIP-44 warns that not all wallets use compatible derivation schemes, so test your derived addresses before you rely on them.

Deriving Keys in Practice

In practice, you need only a couple of pieces of data to derive a near-infinite number of new addresses:

- The master key

- The integer representing the appropriate account

- The index value from which to start counting

Both account number and index number start from 0, so if you want simplicity, start with a path of m'/44'/3'/0'/0/0 and increment only the index value.

If you're associating addresses with transaction marker (such as an arcade machine, a customer invoice, or a customer account), you'll probably want to store the derived address with the index value in a persistent store (for example, in a database) so that you can take the appropriate action when you see a confirmed transaction to that address.

How would that code look?

Example Derivation Code

The libdogecoin library provides useful functions to derive new addresses from hardened master private keys, at least in versions 0.1.3 and later.

While you can write C or C++ code to link against the library directly, you might prefer using a language binding, such as Perl's Finance::Libdogecoin (again, at least version 0.1.3), with code as:

```perl
use 5.038;

use Finance::Libdogecoin 'get_derived_hd_address_by_path';

sub derive_address ($key, $index) {
    my $path    = "m/44'/3'/0'/0/$index";
    my $address = get_derived_hd_address_by_path($key, $path, 0);
```

319

```
        return { address => $address, path => $path };
    }
```

Given the master key and the new index to use, this code calls `libdogecoin` to derive a new address and returns the address and the derivation path used to generate it. **Be aware** that it's your responsibility to manage the security and secrecy of your private keys.

In a web context, you might want to build a model object more like this:

```
use Object::Pad;

class Deriver {
    use Crypt::Lite;
    use Finance::Libdogecoin 'get_derived_hd_address_by_path';

    field $key;

    ADJUSTPARAMS( $params ) {
        $key = Crypt::Lite->new->decrypt(
          $params->{master_key},
          $params->{secret}
        );
    }

    method derive_address ($index) {
        my $path    = "m/44'/3'/0'/0/$index";
        my $address =
            get_derived_hd_address_by_path($key, $path, 0);

        return { address => $address, path => $path };
    }
}
```

This code uses `Crypt::Lite` to decrypt the master key. This allows you to store the master key in a file on disk, encrypted with a secret. When you create this object, you can pass the master key and a secret value so that the object will decrypt the master key and store it in this object's `$key` attribute.

All subsequent calls to the `derive_address` method will use the master key without you having to pass anything. When you start this program, you should interactively enter the secret (or otherwise inject it into the program) so that decryption works correctly *and* no one can get to the secret without your permission.

For a further enhancement, you could store the current `index` value in the object itself, such that every call to a method would increment that value and give you a new address.

None of these examples require the use of Perl, of course. The same techniques work in any language which provides `libdogecoin` bindings; even C itself[11].

Understand the Risks

The biggest risk with the `libdogecoin` approach is that, currently, the code that derives new addresses needs access to the master key. You can keep that encrypted as much as possible, but at some point it's unencrypted in memory. Because this is the master key used to derive child keys, anyone with this key could themselves drive new keypairs (both public and private) as well as the related addresses.

Granted, someone with access to your machine could inspect your memory, but a safer approach would be if you could derive new keys and addresses from a *public* key only[12]. In that case, any exposure of that key would compromise the *privacy* of your public keys and addresses but not the security of your private keys.

As always, it's up to you to decide how to balance security, privacy, and convenience. Filling a pool of fresh addresses requires ongoing maintenance, but it lessens exposure of your keys. Generating new addresses on demand is more convenient and needs less maintenance, but it raises the risk of key exposure.

  Tip #99   Program Real Buttons

Before business starts booming in your little arcade, you have to fill in some gaps. You've figured out whether you want to use dedicated addresses per machine, a token payment system, or an à la carte system. You don't want people to put dollar bills, quarters, or other minted coins into your machines–too much to manage, too unhygienic, too easy to get ice cream everywhere.

Now you need a way to turn a Dogecoin payment into a game credit–some way to turn a blockchain transaction into a button push, a switch flip, or a relay click. You have options.

Programmable Relays

What happens when you put a quarter in a pinball machine? You put the coin in a slot. The coin rolls down a ramp, and, if the coin is valid, it hits a switch (connecting two wires and closing a circuit) and falls down into a hopper (releasing the

---

[11] Though writing web applications in C is even more of a flex now than it was for even a programmer as smart as Raph Levien in 1999.

[12] You *can* do this in newer versions of `libdogecoin`; see the documentation for pros and cons.

switch and disconnecting the wires and opening the circuit). The important part here is that the coin toggles a switch.

When your cabinet registers the switch toggle, it adds credits to the machine, which allows you to press the one or two-player button and start the game. Your cabinet doesn't care that a quarter toggled the switch. It only cares about the circuit closing and then opening. You could achieve the same thing by replacing the coin mechanism entirely, jumping the circuit with a little piece of wire[13], or *installing your own button*.

For this to work, assume you're okay adding new hardware to your existing cabinets, you feel fine using wifi to control *something*, you will enjoy doing a little bit of programming and configuration and hardware manipulation, and you can spend a little bit of money. You can change any of those assumptions, but the basic concept remains: you must turn a computer signal to a physical pinball signal.

The easiest way to do this is to use a programmable relay. A programmable relay is essentially a switch controlled by a small computer. A switch, of course, is a device that connects two wires. When it's on, the wires are connected. When it's off, the wires aren't. Connect this relay to the appropriate circuit inside your cabinet and now you can toggle the switch and the machine will act as if you'd inserted a quarter.

Finding a Good Programmable Relay

For the purpose of this exercise, we'll assume that you have access to a programmable relay board such as an ESP8266 4-relay board[14]. The ESP8266 chip is a tiny 32-bit computer with wifi and some programmable capacity. The relay part is four little programmable switches. While you can buy a 1-, 2-, or 8-relay board, hold that thought.

Improving a Programmable Relay

This board and chip have good and reliable open source firmware available. This will help! The Tasmota firmware worked well in your author's experiments[15]. To make this work, you must flash the new firmware onto the chip, which requires you to use a USB-to-serial adapter and a careful hand to connect the correct pins to the correct wires on the board. Read the directions carefully.

---

[13] Don't actually do this; you're liable to zap yourself.

[14] You can find this from many reputable electronics suppliers for less than $5 USD.

[15] See https://tasmota.github.io/docs/Getting-Started/ to begin.

When you flash the board, you'll get an option to connect to the board over wifi to configure it to connect to a different wifi network (provide your credentials) and program the relays to your liking. You'll find many other options to play with, but for now, you only need to toggle the relays.

Programming the Relay

This is where things get more interesting. Think back to your cabinet's coin mechanism. The coin insertion switch closes for a fixed period of time then opens again. You need to emulate that sequence of events–circuit on, wait, then off. Your relay board must receive an event from the network, close the switch, wait for a fixed period of time, then open the switch again.

How do you know how long to wait? You can experiment with different times, but the best approach is to get the service manual for your cabinet and understand the scan times of the circuits. In general, start with a quarter of a second and adjust from there–perhaps up to half a second.

Arcade Controller Scan Matrix

What governs how long the switch must remain closed? It's all about your machine. Your cabinet is doing a lot of things at once. It scans every potential input once every n milliseconds. The capacitors and resistors on the controller board help debounce the switches, to smooth out any spikes and turn one analog event into one digital event. The controller may also have software controls to debounce the values. Careful measurement and timing will give you the answer. Or you can experiment with the given values.

Testing the Program

With the Tasmota firmware on the ESP8266, you can control the relay with MQTT (your best option for a complex configuration) but also simple HTTP (for testing or a simple network). You'll need to write a program or use `curl` to send these requests. Here's a simple example that tells the relay to close for 200 milliseconds, then turns on the relay, then lets it close:

```
$ curl http://192.168.1.10/cm?cmnd=PulseTime1%202
$ curl http://192.168.1.10/cm?cmnd=Power1%20ON
```

In this example, the board has an IP address of 192.168.1.10 and the relay to toggle is relay 1. Change the example appropriately for your board. The `PulseTime` command sets the time the relay will stay on in milliseconds. As you might expect, relay 2 uses the command `PulseTime2` and so forth. Similarly, the `Power` command turns the relay on or off, with the appropriate relay number and the subcommand `ON` or `OFF`.

Remember that the characters %20 represent a space encoded in a URL.

Before you open the inside of your cabinet and start poking around with a multimeter, test your relay board by connecting an LED to the switch and running your program. Make sure your test LED supports the voltage you're sending!

When your LED flashes on and off for the right timing, you're ready to move on! You should be able to send both commands, see the LED light, and then turn off in the right amount of time. From there, your program is working and you can integrate it with your payment receiving webhooks!

What Can You Do With This?

You can use a programmable relay to turn on a light, open a door, drop cat food into a bowl, and do anything you can imagine a switch would control. In terms of your arcade, you can add credits to a machine, sure, but you can also turn *off* the machine entirely, turn on a siren, or do anything else.

A multi-relay board will allow you to control multiple machines from a single board. You'll have to run wires between them somehow, which may mean drilling into your cabinets from the bottom or adding another wire to a harness elsewhere.

Keep in mind a few limitations, however. First, using a wifi-enabled board means you have a wifi network available. Anyone who can get access to that network has the potential to get access to your system and can control your machines. That may not be a security problem for you, but if you also keep your Dogecoin payment system and wallet on the same network, then attackers have *two* interesting systems to exploit.

Second, you're limited by the number of relays on the board as well as any other physical limitations. If all of the cabinets connected to a single board have the same type of switches (all five volt relays, for example), then you have a simpler time connecting them all together. If they differ–as they likely do–then you have to be careful about how you wire things together.

Third, any oddities of the board or its relays will affect your network. For example, this Tasmota firmware will turn on relay 1 when the board first boots[16]. This may or may not be a problem; you may or may not want to work around it. Furthermore, any time the board reboots, you may lose configuration you'd set–so while it's

---

[16]See https://templates.blakadder.com/ESP12F_Relay_X4.html.

appropriate to set the pulse time once on boot, you may want to set it before every pulse. That way, you *know* the relay will perform as expected for every command you send it (and sending two HTTP requests is almost as cheap as sending one).

Finally, the board itself needs power. The board in this example has a lot of power options including wall warts, USB power, and direct-circuit connections, but that's for you to figure out, lest you fry the board or have to spend a lot of time thinking about how to get electricity from the wall to it.

On the other hand, the thought of tapping a friendly Doge icon on your phone and seeing an arcade machine light up, inviting you to play, is cool.

Tip #100   Flip a Switch

You've installed your arcade hardware, you've set up machines with addresses, you've set up your tokens, and you've programmed some relay boards (see Program Real Buttons, pp. 321) that can close temporarily and then open again. In theory, you have a payment system that can turn a Dogecoin purchase into a game of Asteroids.

What's left? Obviously getting people in the doors to play games, but also the final *technical* connection: turning a payment into the Dogecoin Arcade equivalent of someone inserting a quarter.

Event-Based Switch Flipping

Whether you use the approach of one Dogecoin address per machine (each transaction sent to the address adds credits) or one address per customer (each Dogecoin transaction adds credits, and the customer spends them on each machine), the last step is to turn some event into a switch flip. That means sending a signal to each relay board.

Assume you have the boards installed appropriately for your system (see Program Real Buttons, pp. 321).

Connect Your Relays to the Network

The programmable relay boards used as examples in this chapter have network connections. You can use wifi (no need to run wires, but you have to secure the wifi network) or Ethernet (you have to run wires, but the network is more secure). Either way, you need some mechanism to send and receive data to and from the boards over the network.

If you do use wifi, create a private network with a strong password. You don't want someone sniffing around your network and hogging your Ghostbusters pinball machine without paying for things. If you do use Ethernet, make sure you don't leave a cable hanging out somewhere or an open port on a switch or hub available. You don't want someone plugging in a laptop and hogging your Lord of the Rings pinball machine without paying for things.

Associate Relay Addresss with Machines

With each board on the network, you can refer to them by IP address or hostname. The good news about hostname is that you can change the IP address (let the network manage their addresses) and you don't have to remember the details of the machines. The bad news is that this configuration is a little bit more work— either set up a DNS server or keep the hostnames in a file somewhere.

Either way, you have to know which relay goes to which machine. Furthermore, if you use one board to manage multiple machines, you have to know which relay of each board goes to each machine. While you may still have a single relay per machine, you'll allow yourself a little bit of future-proofing by associating the board, the relay on the board, and the machine together.

In SQL, you might have a table like:

```
CREATE TABLE addresses_to_relays (
  machine CHARACTER(64) NOT NULL,
  dns_name TEXT NOT NULL,
  relay INTEGER NOT NULL,
  is_active BOOLEAN NOT NULL DEFAULT false
);
```

In this data model, `relay` refers to a specific relay on each board and `machine` gives you a human-readable way to refer to a specific machine. The most important change here is that a machine can be associated with multiple relays[17].

Send a Signal to the Relay Board

Now to the final part: to simulate a button press, you send a signal over the network to the relay board. If you've programmed the board appropriately, it knows to close the relay for a short period of time and open it again (simulating a quarter triggering the coin mech). You'll have to send an HTTP message to the appropriate board (you know its DNS name) *containing* the appropriate relay number and then the magic happens.

You can use `curl` to send the message. It will look something like this:

---

[17]For the SQL-adventurous, you can create a unique index on the first three columns as you prefer.

```
curl http://lotr-pinball-board.local/cm?cmnd=Power1%20On
```

...where `lotr-pinball.local` is the DNS name of the relay board and 1 is the number of the relay connected to the pinball machine's Player 1 coin mech.

Make sure you have your board installed, your relay connected, your network configured, and your machine powered on. If everything is ready to go, you'll hear a Balrog roar.

Automate the Process

Of course you don't want to type `curl` commands every time someone pays for a play. You'll need to set up the appropriate webhook. The good news is that the POST command *is* that webhook. In fact, you can use the same script to send the same message to different relays in one of two ways:

- look up the machine name, board DNS name, and relay number on each execution (hide everything behind a single webhook and double-dispatch)

- generate and register webhooks on each change of the `addresses_to_-relays` table (single dispatch)

You can generate a list of webhook URLs with SQL something like:

```
SELECT
    'http://'
    || dns_name
    || '/cm?cmnd=Power'
    || relay
    || '%20On
  FROM addresses_to_relays
  WHERE is_active;
```

Register these webhooks appropriately and the entire system should work, start to finish.

Alternate Approaches

This approach uses HTTP GET requests over HTTP. You're probably not exposing any secret or sensitive information over the network, but if you prefer to use HTTPS everywhere, you can configure this on certain relay boards.

You may prefer not to use HTTP overall. The example relay boards used in this chapter also support MQTT, which requires a little more configuration but gives you a lot less administration in the proper circumstances. Play around and see what you can discover!

Understand the Risks

Every component and connection you add to the system is a potential point of attack and failure. If your network goes down, a lot of people will sit around waiting for a chance to save the city. If your relay board makes a popping sound and the magic smoke escapes, you'll have to replace it. If your DNS system resets and IP addresses switch around, you'll hear Spy Hunter rev up when someone wants to play Turtles in Time.

Regular diagnostics and checks are important. You can automate things like health checks and, given the `is_active` column in the example SQL tables, take a machine/board/relay combination out of the system. To make the entire system more robust, you can use this as a status to disable the ability of users to add credits to a machine (though you can't prevent someone from sending a Dogecoin transaction to an address).

Also remember that every new configuration option added to the system adds flexibility as well as management overhead. Avoiding webhook double-dispatch simplifies the system but makes it more difficult to unregister a webhook when you need to maintain a machine or connection.

Depending on how seriously you take this business, having extra, easily swappable, hardware on hand can save you a lot of time and trouble. You don't want to be reprogramming your relay board while customers are lining up and tapping their feet.

You want people to play games and have fun. The fun you had building the system may not be the fun your customers want–but if you do things well, you can maximize the enjoyment everyone gets by removing friction, promoting safety, and building new and interesting things with your favorite dog-themed programmable money.

  Tip #101   Install Programmable Buttons

This is it: the moment of truth. Think back to everything that's built up to this so far. You've learned the basics of cryptography and the ins and outs of running a Dogecoin node. You've explored sending commands to and receiving data from the network.

You've sent and received transactions. You've generated and used and secured addresses. You've thought about payment systems and wallet security. You've played games and explored puzzles.

Along the way you've read about making actual machines do actual work (see

328

Program Real Buttons, pp. 321 and Flip a Switch, pp. 325) and you've done things when your node saw things happen on the network (see Take Actions on New Blocks, pp. 83).

Now it's time to put what you've learned together and wire up a real actual pinball machine to the Dogecoin network.

What You Need

It's okay if you don't have a big, heavy, expensive pinball machine lying around. Anything that can be controlled with a simple electric switch will do: coffee machine, music player, lava lamp, flashing sign, or anything else.

This tip assumes you have:

- programmed a relay board to control a switch from a network command

- basic electronic parts, such as wires, Molex connectors, alligator clips, and a multimeter

- access to basic knowledge about electricity and circuits, yourself or a friend

- a Core node or other Dogecoin wallet connected to the network

Connecting the Board to the Machine

For this example, your author used a Lord of the Rings Stern Pinball machine[18]. This machine has two coin slots and, importantly, an expansion connector to add a tournament printer.

On your machine, you may have a different configuration, such as a button to simulate adding a coin, a bill acceptor for paper money, or a card reader. Similarly, you may have a power connector intended to supply another accessory or no power at all.

This is important: you need to connect to the machine in at least one, perhaps two ways. First, to toggle the switch which signals to the machine that someone has inserted a credit (simulating a coin, card swipe, or bill insertion). Second, to power the programmable relay board.

You don't *need* to do the second, though it may be more convenient, but you do need to power the board somehow. The nice part about powering the board from inside the machine is that the board is neither visible nor externally accessible; you can lock the coin door to keep people out.

---

[18]To learn more, visit https://sternpinball.com/game/the-lord-of-the-rings/ and https://pinside.com/pinball/machine/lord-of-the-rings.

Getting Power

On the inside right of the LoTR machine, you can find a three-pin Molex connector. This provides a ground wire, a +5V wire, and another wire you can ignore. If your board runs from 5 volts, you can attach two wires to a Molex connector to power the board.

> ### If You Know You Know
>
> Be careful doing this. If you connect the wrong wires, you can damage either or both the board and the pinball machine. Check and double-check the voltage and your connections.
>
> This tip will explain the concepts and show diagrams, but it won't tell you *exactly* how to do things, because you're much safer working with someone who has practical experience doing things like this.

If all goes well, you can power down the machine, unplug the machine, connect the board connector to the machine connector, plugin in the machine, turn on the machine, and see your board light up (then see it on the network).

With power to the board, try sending a signal to the board to toggle the relay. With your multimeter connected to the correct relay terminals, you should see a voltage toggle to the tune of 5 volts. If you don't, check your connections and debug your settings[19].

Connecting to the Switch

In the LoTR machine, the right coin slot is easy to access. It has one blade connector and one soldered connector. When someone inserts a coin, a mechanical lever closes the circuit between the two connectors briefly, and the machine registers a credit.

To simulate this, connect one wire from the relay board to the blade connector and another to the soldered connector. For the specific ESP board, use the wire from the Normally Open (NO) relay terminal to the blade connector and the wire from the Normally Closed (NC) terminal to the soldered connector. If you're testing this now, you can use an alligator clip for the soldered connector instead, if you're

---

[19] Your author managed to deprogram his board trying to get the power connection right, so be prepared to reprogram anything you need there.

very careful not to short anything else. Otherwise, a Scotchlok connector might be better.

Similarly, for the blade connector, consider using a pigtail where you can insert the wire from the relay board *between* the existing connections.

Testing this way allows you to get the figurative bugs out of the system before making permanent modifications to your machine.

Customizing Your System

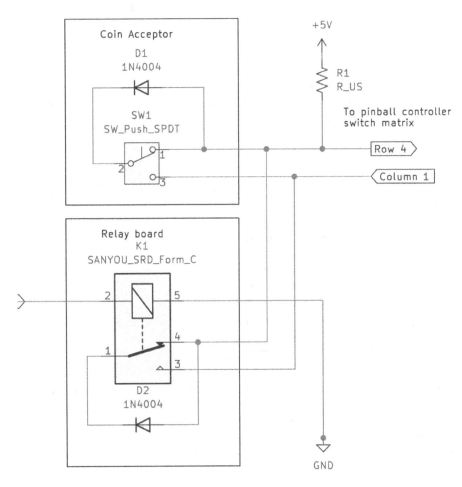

Figure 10.1: LoTR Programmable Relay Schematic

You can take this a lot further, perhaps sharing one relay between multiple machines (in that case, you have to run wires from a central board to all machines, but you only need to power the board once). You can also disable the coin slot or payment mechanism if you like–or even remove it, if you prefer.

See LoTR Programmable Relay Schematic, pp. 331 for a schematic of the relay connection specific to this Stern machine. You can adapt this to your machine by examining its service manual very carefully. In particular, in this schematic, note the 1N4004 diode connected to the relay which matches a similar diode on the machine's coin mechanism.

You may need to add something similar for safety (and to match scan timings of the machine).

With the board connected to the machine for power and the relay connected to the coin slot, now you can send a command to toggle the switch. If all goes well, you'll see the machine register a credit.

If not, check that the relay toggling is the one you intend to toggle, that the connections are secure and not shorted, and that the delay on the toggle is long enough.

Understand the Risks

Whew. All it took to get a quarter's worth of pinball is an entire book full of programmable dog money, but you made it.

The main risk of doing something cool like this is that you'll spend a lot of time and money collecting obscure hardware and then say "Wouldn't it be amazing if I installed a webcam and put a QR code online so people could pay one Dogecoin to see their pinball credits register?" All of the pieces here and in this chapter and in this book can be rearranged and recombined in many ways, some of them not invented yet. Maybe you'll be the one to invent them.

In more seriousness, the risks of doing any electrical work inside a large device like a pinball machine are electrocution, fire, and damage to either device. Again, review the service manual, consult a knowledgeable friend, and test and re-test with your multimeter. This goes whether you're connecting arcade hardware, multimedia equipment, or anything else.

On the other hand, the rewards are pretty cool, and there's no satisfaction quite like taking what you've learned and making something tactile with it–especially when you start with what was supposed to be a joke and now has become something useful.

Index

```
          DOON
         LYd   GO
         Ooo      D
         Enl      V
         EygoR    od Y
         DeveA    r YyD
         OdaydO   oNLoY
         GnlyO    go  OD
         EodevV      ER
         YDeryA        Y
         DdOaydO        N
         LooYGO  O     D
       EV  nlygoE RYDA
       Y       odevD
     O              O
     N              L
     Y              G
     O              O
   D                E
 e  Vr        E     R
 y  Yd D      AY    D
ay Odo ONLY GO      O
 onlD E    V  E  yR
     YDAY  DOO NL YG
```

www.ingramcontent.com/pod-product-compliance
Lightning Source LLC
LaVergne TN
LVHW051428050326
832903LV00030BD/2978